Advances in Gravity Concentration

Edited by R.Q. Honaker and W.R. Forrest

Published by the
Society for Mining, Metallurgy, and Exploration, Inc.

Society for Mining, Metallurgy, and Exploration, Inc. (SME)
8307 Shaffer Parkway
Littleton, Colorado, USA 80127
(303) 973-9550 / (800) 763-3132
www.smenet.org

SME advances the worldwide mining and minerals community through information exchange and professional development. SME is the world's largest association of minerals professionals.

Copyright © 2003 Society for Mining, Metallurgy, and Exploration, Inc.

All Rights Reserved. Printed in the United States of America.

On the Cover
   Top: An industrial circuit employing the Knelson Concentrator for gravity separations on ultrafine particles. A mechanically applied rotation of a fluidized drum supplies the necessary enhanced gravitational field for effective separations.
   Left: Spiral concentrators are used worldwide to beneficiate ferrous and nonferrous minerals, energy and industrial minerals, and to produce mineral derived chemicals.
   Right: The industrial sand processing plant supplied by Outokumpu Technology to St. Gobain, relies on Carpco® spirals to produce sand for glass production at this plant near Chennai in Southern India.

Information contained in this work has been obtained by SME, Inc. from sources believed to be reliable. However, neither SME nor its authors guarantee the accuracy or completeness of any information published herein, and neither SME nor its authors shall be responsible for any errors, omissions, or damages arising out of use of this information. This work is published with the understanding that SME and its authors are supplying information but are not attempting to render engineering or other professional services.

No part of this publication may be reproduced, stored in a retrieval system, or transmitted in any form or by any means, electronic, mechanical, photocopying, recording, or otherwise, without the prior written permission of the publisher. Any statements or views presented here are those of the authors and are not necessarily those of SME. The mention of trade names for commercial products does not imply the approval or endorsement of SME.

ISBN 0-87335-227-0

Library of Congress Cataloging-in-Publication Data

Advances in gravity concentration / edited by R.Q. Honaker and W.R. Forrest.
      p. cm.
   Includes bibliographical references and index.
   ISBN 0-87335-227-0
      1. Gravity concentrators. I. Honaker, R. Q. (Ricky Quay), 1963– II. Forrest, W.R. (William Richard), 1961–

TN520.A38 2003
622'.751--dc21

2002044775

Contents

PREFACE v

Density Separations: Are We Really Making Use of Existing Process Engineering Knowledge?
Gerald H. Luttrell **1**

SECTION 1   **GRAVITY CONCENTRATION FUNDAMENTALS** **17**

On the Phenomena of Hindered Settling in Liquid Fluidized Beds
K.P. Galvin **19**

Modeling of Hindered-settling Column Separations
Bruce H. Kim, Mark S. Klima, and Heechan Cho **39**

Dense Medium Rheology and Its Effect on Dense Medium Separation
Janusz S. Laskowski **55**

Methodology for Performance Characterization of Gravity Concentrators
M. Nombe, J. Yingling, and R. Honaker **71**

SECTION 2   **COAL-BASED GRAVITY SEPARATIONS** **79**

Optimum Cutpoints for Heavy Medium Separations
G.H. Luttrell, C.J. Barbee, and F.L. Stanley **81**

Operating Characteristics of Water-only Cyclone/Spiral Circuits Cleaning Fine Coal
Peter Bethell and Robert G. Moorhead **93**

Comparing a Two-stage Spiral to Two Stages of Spirals for Fine Coal Preparation
Peter J. Bethell and Barbara J. Arnold **107**

Advances in Teeter-bed Technology for Coal Cleaning Applications
Jaisen N. Kohmuench, Michael J. Mankosa, and Rick Q. Honaker **115**

Innovations in Fine Coal Density Separations
R.Q. Honaker and A.V. Ozsever **125**

SECTION 3   **NON-COAL GRAVITY SEPARATIONS**   **141**

      Heavy Media Separation (HMS) Revisited
Roshan B. Bhappu and John D. Hightower   **143**

      Recovery of Gold Carriers at the Granny Smith Mine Using Kelsey Jigs J1800
G. Butcher and A.R. Laplante   **155**

      Applications of the HydroFloat Air-assisted Gravity Separator
Michael J. Mankosa, Jaisen N. Kohmuench, Graeme Strathdee, and Gerald H. Luttrell   **165**

      Advances in the Application of Spiral Concentrators for Production of Glass Sand
Steve Hearn and Jim Sadowski   **179**

INDEX   **189**

Preface

Gravity-based separators have been a mainstay of the mineral and coal processing industries for centuries. Georgius Agricola described a sixteenth-century application of gravity separation for tin and copper recovery in the book entitled *De Re Metallica* (1556). Since that time, advances made during the industrial revolution and the more recent development of magnetic separation technologies (employed for dense-medium recovery) have contributed to many of the processes employed today. But gravity-based separators still remain the prominent means of producing concentrates from coal, iron ore, rare earths, industrial minerals, tin ores, and tungsten ores. Also, the recycling industry prominently uses gravity separators to recover the various materials from previously disposed waste.

Due to the complexity and cost of flotation systems, greater attention is being given to reduce the lower particle size limitation for effective gravity-based separators. An additional benefit of gravity-based separations is the potential to improve effectiveness in treating mixed-phase particles as compared to separations achieved by froth flotation. Efficient, high-capacity gravity separations now can be achieved for particles as small as 10 microns using gravity-based units that provide centrifugal forces as high as 300 times the natural gravitational force. Research equipment has become more sophisticated and provides opportunities for in-situ studies of processes, resulting in new fundamental theories and control schemes. The computer age and the development of robust on-line analyzers have allowed full automation and on-line optimization of gravity-based systems. Advanced technologies developed in other professional fields, such as the medical sciences and inventory control, are being employed to provide on-line washability analysis, thereby improving operation efficiency. However, during the time period of these developments, few opportunities have existed to concentrate the efforts in a national or international platform where practitioners and researchers can present and discuss the current developments and future trends. This publication provides a review of the state-of-the-art in gravity concentration and their associated applications.

The book is subdivided into three major areas: fundamentals, coal applications, and non-coal applications. The fundamentals section details developments in the knowledge of particle characterization, particle-settling kinetics, slurry rheology, and overall process modeling. Novel technological and circuitry advances will be reviewed in both coal and non-coal applications. Technologies incorporating other physical forces, such as those associated with surface chemistry properties, are discussed and their relative efficiencies provided.

The proceedings editors and section organizers greatly appreciate the contributions of all authors and co-authors. In addition, sincere gratitude is expressed to Mr. John Mansanti, SME MPD chairman, for his support of the symposium; Ms. Tara Davis, SME program manager; and Ms. Jane Olivier, SME manager of book publishing, for their help and guidance in the development of the symposium and this proceedings.

Density Separations: Are We Really Making Use of Existing Process Engineering Knowledge?

Gerald H. Luttrell[*]

Considerable process engineering knowledge has been developed regarding the design and operation of density separators and circuits. Two of the most important of these are the incremental quality concept advocated by Abbott (1982) and the circuit analysis principle described by Meloy (1983). These versatile concepts are equally important to engineers involved in flowsheet design, equipment manufacturers involved in process development, and personnel involved in the daily operation of industrial plants. Unfortunately, experience suggests that these groups are either unaware of these concepts or fail to recognize how they may be applied to solve real world problems. In light of this shortcoming, this article provides a brief overview of these important process engineering concepts and provides practical examples to illustrate how they can be applied to the design, control, and optimization of density separators and circuits.

INTRODUCTION

Density-based separation processes are commonly used to upgrade a variety of materials including coal, tin, iron ore, and heavy mineral sands. These processes are often incorporated into parallel and/or multistage processing circuits at industrial sites. Parallel circuits are required because of particle size limitations and throughput restrictions associated with commercial density separators. For example, coal preparation plants are forced to include three or more parallel circuits as part of their basic flowsheet because no unit operation currently exists that can upgrade a very wide range of particle sizes. Likewise, some mineral beneficiation plants are forced to distribute feed slurry to dozens or more spiral separators because of feed throughput restrictions. Several stages of cleaning and scavenging in series are also commonly required in mineral plants to overcome inherent inefficiencies created by the random misplacement and/or bypass of particles in density-based separation processes.

---

[*] Dept. of Mining and Minerals Engineering, Virginia Polytechnic Institute and State Universtiy, Blacksburg, Va.

Recent field studies suggest that many of the industrial processing plants that make use of density separators have the potential to significantly improve metallurgical performance through the proper application of existing process engineering expertise. One of the most important process engineering principles is the *incremental quality concept*. This optimization concept provides basic insight into how parallel processing units must be operated in order to maximize product recovery. This concept also has important implications related to product blending practices and plant control strategies. Another key process engineering principle is that of *linear circuit analysis*. This analysis technique provides generic rules that govern how multiple unit operations should be arranged within a circuit to maximize separation efficiency. This powerful tool can be used to quickly evaluate the effectiveness of a variety of circuit configurations and can substantially minimize the need for trial-and-error experimental testing. Unfortunately, industrial operators and plant designers often overlook these two important process engineering principles. The objective of this article is (i) to provide a general review of the incremental quality concept and linear circuit analysis and (ii) to present industrial examples that demonstrate how these principles may impact the profitability of industrial operations.

INCREMENTAL QUALITY CONCEPT

The optimization of separation processes based on the concept of constant incremental quality has long been recognized in the technical literature (Mayer, 1950; Dell, 1956). This concept states that a blended product from one or more parallel circuits that is constrained by an upper limit on concentrate grade will provide maximum yield only when all units are operated at the same incremental quality. In layman terms, incremental quality is the effective grade of the last material added to the concentrate (or removed from the tailings) when the cutpoint to the process is increased by an infinitesimal amount. This optimization principle is valid for any number of parallel circuits and is independent of the particle size and liberation characteristics of the feed streams. It can also be shown that plant profitability is maximized when constant incremental quality is maintained regardless of the operating costs of the different separation processes (Abbott, 1982).

A mathematical proof of the incremental quality concept has been provided elsewhere in this volume (Luttrell et al., 2003). However, the fundamental basis for this concept can be demonstrated using the simple illustrations provided in Figure 1. This series of diagrams shows two feed streams (A and B) comprised of various amounts of valuable particles (dark colored material), gangue particles (light colored material), and middlings. The composition of Stream A makes it relatively easy to upgrade since most of the particles are well liberated. In contrast, Stream B is difficult to upgrade due to the presence of a large percentage of middlings (locked) particles. Figure 1(a) shows the result that is obtained when the separation is conducted to provide a target grade of 75% for both streams. For Stream A, the cutpoint is increased from left to right until a total of 9 blocks of dark (valuable) material is recovered out of a total of 12 blocks. This operating point produces a yield of 75% at the desired target grade of 75% (i.e., 6.75/9 = 75%). All but the pure gangue particles must be recovered to reach the target grade when operating under this condition. For Stream B, a yield of only 53.8% (i.e., 7/13 = 53.8%) can be realized before the target grade of 75% (i.e., 5.25/7 = 75%) is exceeded. The lower yield is due to the poorer liberation characteristics of this feed stream. The concentrate from these two streams is blended back together to produce a combined yield of 64% at the desired target grade of 75%. At first inspection, this appears to be an acceptable result since the target grade is indeed met. However, a closer examination shows that particles containing up to 75% gangue reported to concentrate when treating the Stream A, while

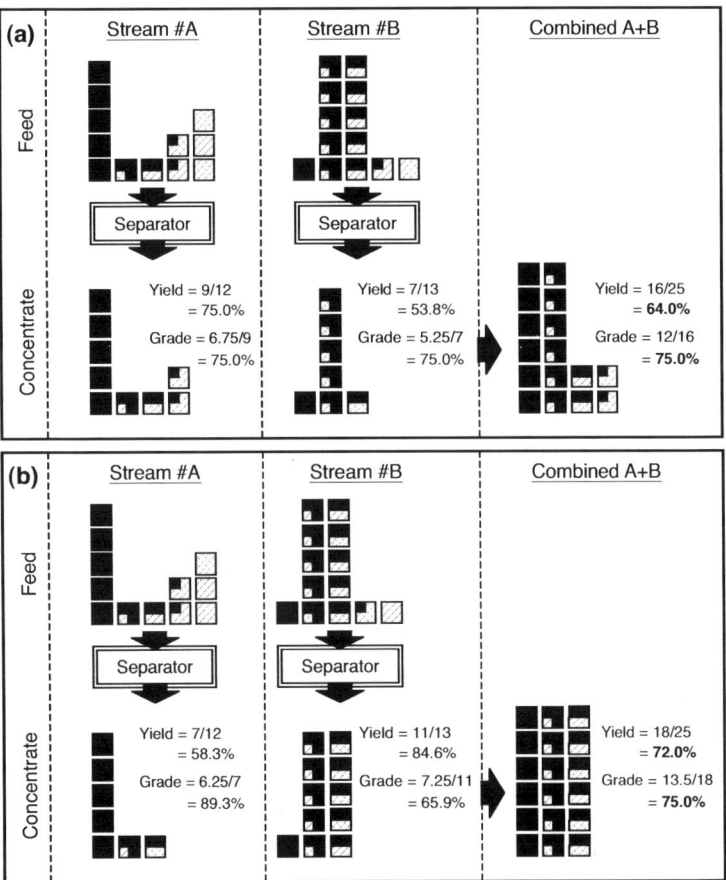

FIGURE 1 Comparison of concentrate yields and grades for parallel feed streams obtained by operating at (a) constant cumulative grade and (b) constant incremental grade

at the same time many particles containing just 50% gangue were discarded when treating Stream B. This result is obviously not optimum since higher quality particles where sacrificed to make room for poorer quality particles in the concentrate.

The incremental quality concept dictates that maximum yield can only be obtained when the two feed streams are treated to provide the same incremental grade, i.e., the quality of the last block recovered from each feed stream must be of the same grade. To test this principle, the feed streams shown in Figure 1(b) are each separated so that the last block recovered in both cases contained 50% valuable material. Operation under this condition increases the concentrate grade for Stream A from 75.0% to 89.3% and decreases the grade for Stream B from 75.0% to 65.9%. However, when the two streams are blended together, the combined concentrate still meets the required target grade of 75%. More importantly, this new set of operating points provides a higher combined yield (i.e., 72% versus 64%). In practice, it is easy to be fooled into accepting the lower yield because of the presence of multiple feed streams. The optimum result would be intuitive, however, if the feed streams are blended together prior to treatment. In this case, the pure particles would be recovered first, then the 75% pure particles, then the 50% pure particles, and so on until no additional particles could be taken without exceeding the target grade. This sequence will always provide the maximum recovery of valuable material at a given grade.

Two important extensions are necessary to fully take advantage of the incremental quality concept. First, the concept may be applied to any particle system, even if the particles present in the feed are completely liberated. Although the concept was originally developed for separations involving systems that included true middlings particles, a separation that involves only two pure minerals would still generate products with "effective" incremental grades because of inefficiencies in the separation process. In cases such as this, the separation is still optimized when the same incremental grade is maintained in all producing circuits. As such, the incremental quality concept can be universally applied to nearly all types of particle separations. Second, incremental quality can often be related to other measurable parameters that are inherent to a given mineral system. This is extremely important since the direct on-line measurement of incremental quality is typically not possible. As an example of this relationship, consider the various types of particles shown previously in Figure 1. The percentages of dark and light material in each block directly control the individual grade of that particular block. If the dark and light components represent high and low density minerals, then these percentages also control the overall density of the block. In fact, the grade of valuable mineral in an individual particle can be mathematically shown to be directly proportional to the reciprocal of the particle density for a two-component system (Anon., 1966). This correlation holds true across the entire particle size range (e.g., a 1 mm particle and a 0.1 mm particle that both contain 50% of a high density component will both have the same grade). Because of this relationship, incremental quality can often be held constant by maintaining the same specific gravity cutpoint in all density separation circuits. Operation under this condition will provide maximum yield at any desired value of cumulative product grade. This statement is true regardless of the size distribution or liberation characteristics of the two-component feed, provided that (i) the densities of the host minerals that make up the particles are not highly variable and (ii) very efficient separations are maintained in each circuit. If the separation is less than ideal, minor corrections are necessary to determine the actual specific gravity setpoints required to provide a given incremental grade (Armstrong and Whitmore, 1982; Rong and Lyman, 1985). Mathematical simulations can be carried out in such cases to identify suitable cutpoints. Methods for estimating these corrections have been discussed elsewhere in this volume for heavy medium separations of run-of-mine coal (Luttrell et al., 2003).

LINEAR CIRCUIT ANALYSIS

Linear circuit analysis was originally developed by Meloy (1983) as a global technique for evaluating the effectiveness of different configurations of unit operations in mineral and coal processing circuits. The underlying principles of this technique can be best illustrated by means of a series of simple examples. Consider the single-stage process shown in Figure 2(a). The concentrate-to-feed ratio (C/F) can be expressed in terms of a dimensionless probability function (P) that selects particles to report to concentrate based on one or more physical properties. For density-based separations, P can be empirically estimated from one of several S-shaped transition (or step) functions such as the one shown in Figure 3. This curve shows the probability function for different values of the normalized specific gravity ratio (Z) given by SG/SG_{50}. The specific gravity cutpoint (SG_{50}) is represented by a value of Z = 1 at which P = 0.5. Since a steeper curve represents a sharper separation, the slope of the partition curve evaluated at Z = 1 can be used as a

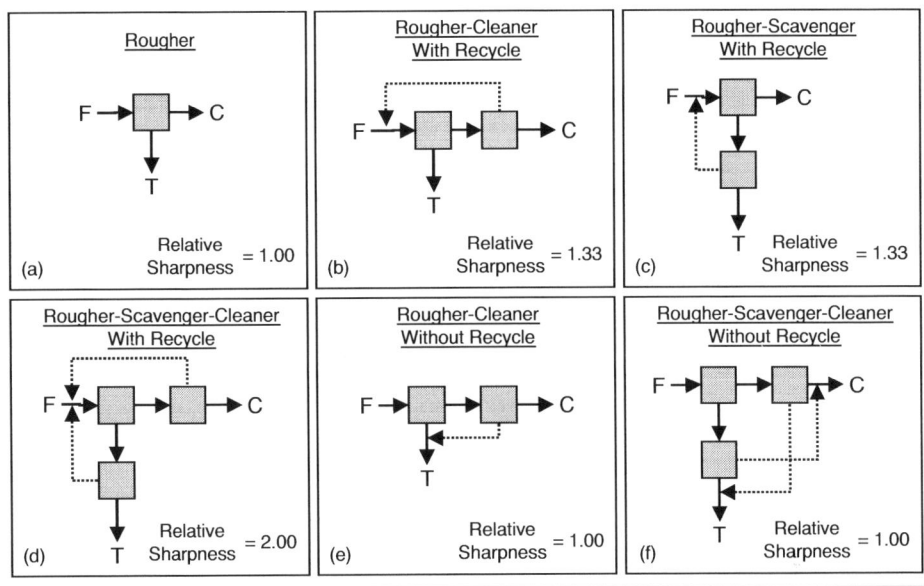

FIGURE 2 Comparison of relative sharpness of separation predicted by circuit analysis for several different circuit configurations

relative indicator of the sharpness of the separation. The slope can be mathematically obtained by taking the derivative of the concentrate-to-feed ratio at $Z = 1$. This gives:

$$\frac{\partial(C/F)}{\partial Z} = \frac{\partial P}{\partial Z} \qquad \text{(EQ 1)}$$

Now consider a similar analysis of a two-stage rougher-cleaner circuit shown in Figure 2(b). In this case, the internal circuit feed (F') to the rougher unit is given by:

$$F' = F + M \qquad \text{(EQ 2)}$$

in which F is the overall feed to the combined circuit and M is the middlings recycle stream from the cleaner unit. The mass of particles of a given property reporting to either the concentrate (C) or middling (M) streams can be calculated from:

$$C = (P_2)(P_1)F' \qquad \text{(EQ 3)}$$

$$M = (1 - P_2)(P_1)F' \qquad \text{(EQ 4)}$$

By simple algebraic substitution, the concentrate-to-feed ratio (C/F) for this circuit can be calculated as:

$$\frac{C}{F} = \frac{P_2 P_1}{1 - P_1 + P_1 P_2} \qquad \text{(EQ 5)}$$

If the probability function (P) is assumed to be the same for all units, then the sharpness of separation for the combined two-stage circuit can again be obtained taking the derivative of the concentrate-to-feed ratio. By noting that $P = 0.5$ at $Z = 1$, the following relationship may be obtained:

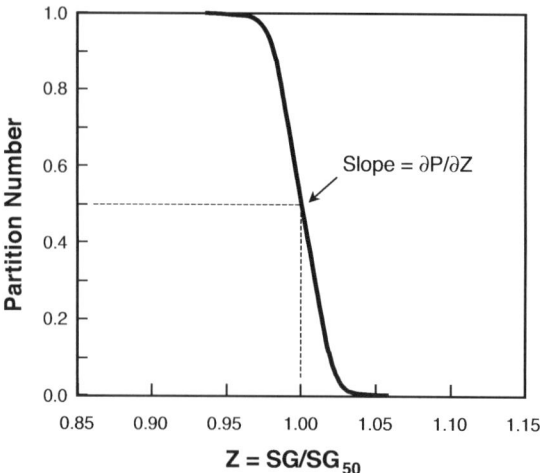

FIGURE 3 Probability distribution function for a generic density-based separation

$$\frac{\partial(C/F)}{\partial Z} = \left(\frac{2P - P^2}{(P^2 - P + 1)^2}\right) = \frac{\partial P}{\partial Z} = 1.33\frac{\partial P}{\partial Z} \quad \text{(EQ 6)}$$

A comparison of Equations 1 and 6 indicate that the sharpness of separation for a rougher-cleaner circuit is theoretically 1.33 times higher than that for a single-stage circuit. Using this same approach, the relative sharpness of separation of other circuit configurations can also be determined. The standard rougher-cleaner configuration (Figure 2b) and rougher-scavenger configuration (Figure 2c) each have a relative separation sharpness that is 1.33 times greater than the single-stage process. The rougher-scavenger-cleaner configuration (Figure 2d) incorporating three total stages has a separation sharpness that is twice that of the single-stage process. Surprisingly, some multi-stage circuits have no better sharpness of separation than a single-stage process (Figures 2e and 2f).

Circuit analysis is obviously an important tool for evaluating different arrangements of unit operations within a processing flowsheet. This technique does not provide specific information regarding product yield and grade that commercial simulation software may provide. However, this technique accurately calculates the same relative numerical efficiencies as commercial packages without the need to have detailed experimental characterization data for the feed streams or empirical partition curves for the unit operations. Some of the important observations that can be derived from circuit analysis are as follows.

- After the primary stage of separation, none of the products generated in subsequent stages should be allowed to cross between the cleaner or scavenger branches of the circuit. Any streams that pass between these branches will ultimately reduce the sharpness of the separation.
- The only configurations inherently capable of improving the sharpness of the separation are those with product streams that are recycled back to previous stages of processing. The recycle streams significantly sharpen the partition curve by providing additional opportunities for misplaced particles to report to the proper steam.

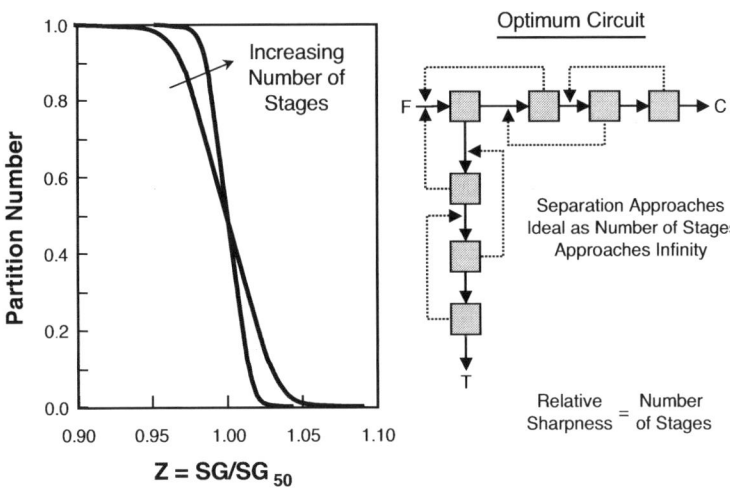

FIGURE 4 Effect of number of cleaning and scavenging stages on the performance of an optimum circuit layout

- When designing a multi-stage circuit, the sharpness separation is obtained when the most efficient process is placed ahead of all other processes as the primary rougher. The order of processing units in the scavenger and cleaner branches does not affect the sharpness of the separation provided that the process streams between different branches are not mixed. Additional unit operations that branch off from the main scavenger and cleaner branches also do not improve the sharpness of the separation.
- The sharpness of separation increases as the number of unit operations down the cleaner and scavenger branches of the circuit increases. This is true provided that the secondary concentrate streams in the scavenger branch and secondary tailings streams in the cleaner branch are fed back to the previous separator in that branch. The performance of a rougher-scavenger-cleaner circuit with recycle streams approaches that of an ideal separator as the number of units in each branch approach infinity (see Figure 4).

INDUSTRIAL APPLICATIONS

Selection of Heavy Medium Cutpoints

Modern coal preparation plants commonly use density-based separators to remove high-density inorganic matter (rock) from low-density carbonaceous matter (coal). The selection of optimum specific gravity cutpoints in these circuits can have a dramatic impact on profitability. Consider the circuit shown in Figure 5 consisting of a heavy medium bath and heavy medium cyclones. The bath treats the plus 10 mm fraction, which represents 30% of the material treated by the heavy medium circuits. The cyclones treat the remaining 70% of the feed material (i.e., 10 × 1 mm fraction). The heavy medium plant is constrained by an upper limit on dry ash of 7% in this particular case. Projected cleanability data for each of these circuits was obtained by numerical partition simulations and are summarized in Table 1.

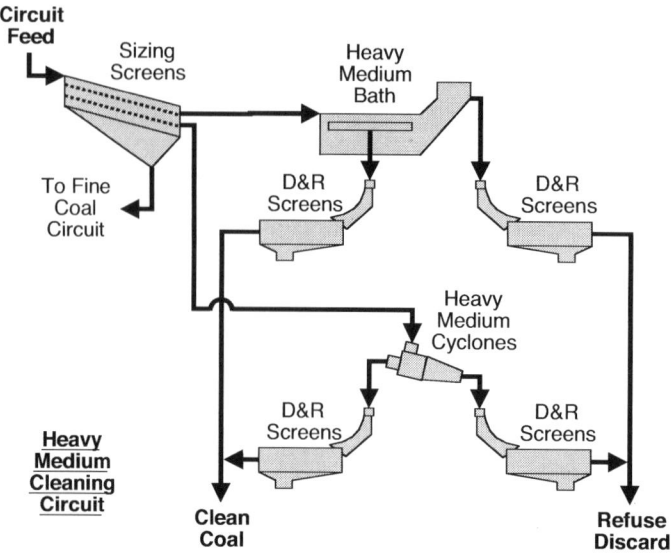

FIGURE 5 Simplified flowsheet for generic coal cleaning plant incorporating a heavy medium bath and cyclones

TABLE 1 Projected cleanability of coals from parallel heavy medium circuits

SG	Heavy Medium Bath		Heavy Medium Cyclones	
	Mass (%)	Ash (%)	Mass (%)	Ash (%)
1.30	11.36	3.86	59.17	3.54
1.35	38.84	7.00	65.27	3.92
1.40	55.10	9.11	68.97	4.41
1.45	58.92	9.83	71.27	4.87
1.50	61.37	10.46	73.15	5.35
1.55	62.63	10.81	74.17	5.62
1.60	63.40	11.13	75.05	6.00
1.65	63.65	11.23	75.56	6.21
1.70	63.96	11.37	76.17	6.48
1.80	64.36	11.58	77.21	7.00
1.90	64.99	11.98	78.21	7.56
2.00	65.53	12.39	79.16	8.20
Feed	100.00	38.64	100.00	24.67

The plant was originally configured to operate with specific gravity cutpoints so that both heavy medium circuits generate clean coal products containing 7% ash. This selection of cutpoints seems obvious since it provides a clean coal having the required product specification. As shown in Table 1, the cutpoint of the heavy medium bath must be set at 1.35 SG in order to produce a 7% ash. This operating point provides a corresponding clean coal yield of 38.84%. Likewise, the cutpoint for the heavy medium cyclones must be set at 1.80 SG to achieve a 7% ash. This value corresponds to a clean coal yield of 77.21%.

The higher yield in the cyclone circuit is due to the superior liberation of the 10 × 1 mm fraction treated by the cyclones compared to the plus 10 mm material treated by the bath. When the products from the bath and cyclone circuits are combined, the overall plant yield is 65.70% (i.e., 0.30 × 38.84 + 0.70 × 77.21 = 65.70). The ash content obviously meets the contract target since both circuits produced a 7% ash clean coal.

For many plant operators, the operating point suggested above seems very reasonable. However, the application of processing engineering knowledge suggests that substantial improvements in the performance of these two circuits should be possible. In fact, the total plant yield calculated above cannot be optimum since the incremental quality concept dictates that the yield is maximized only when the processes are operated at the same incremental ash. This normally dictates that specific gravity cutpoints be held constant for efficient processes such as heavy medium separators. Therefore, let us consider the case where both circuits are set to operate at identical specific gravity cutpoints of 1.55 SG. The bath produces a 62.63% yield at 10.81% ash, while the cyclones produce a 74.17% yield at 5.62% ash. When combined, these products provide a substantially higher plant yield of 70.71% (i.e., 0.30 × 62.63 + 0.70 × 74.17 = 70.71). This operating point still provides an acceptable combined product ash of 7% (i.e., [0.30 × 62.63 × 10.81 + 0.70 × 74.17 × 5.62]/ 70.71 = 7.00). The net difference in plant yield is 5.01% (70.71 – 65.70 = 5.01%). If the heavy medium circuits treat a total of 750 tph of run-of-mine feed, then this modification to the plant operating practice represents an increase in marketable production of about 37.6 tph (750 tph × 5.01% = 37.6 tph). The higher clean coal tonnage would account for an increase in net revenues of more than $5 million annually assuming a $25/ton clean coal sales price and 5,500 operating hours per year (37.6 tph × $25/ton × 5,500 hr/yr = $5.17 MM). Unfortunately, a recent field survey conducted by the author suggests that nearly two-thirds of eastern coal preparation plants do not attempt to operate their plants at a constant incremental quality. These poor operating practices continue despite the fact that this optimization concept has been known for more than half a century.

Evaluation of Plant Control Alternatives

Some of the important applications of the incremental quality concept are not entirely obvious. For example, consider a plant where the feedstock is subject to significant variations in terms of particle size and grade. These disturbances make it difficult for plant operations to maintain a consistent product quality that is demanded by downstream customers. Factors responsible for these variations include natural fluctuations in the physical properties of the run-of-mine feed or routine changes in production rates from multiple sections or mines that supply ore to the processing facility. Many operations resort to automatic or semi-automatic control systems to help alleviate this problem. In most cases, these systems utilize feedback from one or more on-line sensors to manipulate process cutpoints so as to achieve a stable product grade. Unfortunately, existing process engineering knowledge suggests that this control strategy is less than ideal. As stated earlier, the incremental quality concept requires that all circuits that contribute to a plant concentrate must be operated at the same incremental grade to achieve maximum yield. What is less obvious is that this optimization concept also requires that the same incremental grade be maintained at all points in time throughout the entire duration of a production cycle. A traditional control system that manipulates circuit cutpoints to maintain grade may recover individual particles containing a low amount of valuable mineral at one point in time when the feed contains an abundance of high-grade particles. Then, at a

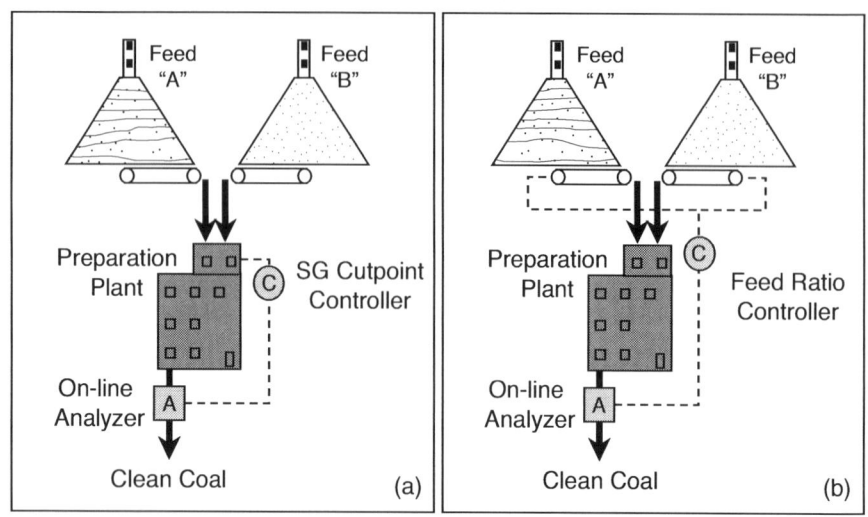

FIGURE 6 Plantwide control systems based on (a) adjustment of specific gravity cutpoints and (b) adjustment of feed rate ratios

later point in time, the same control system will discard individual particles containing a larger amount of valuable mineral when the feed contains fewer high-grade particles. As a result, a plant that continuously raises and lowers cutpoints for grade control purposes will always produce less concentrate than a plant that maintains the same incremental grade. To overcome this problem, the plant should if at all possible operate at all points in time with constant cutpoints in an attempt to maintain a constant incremental quality. This mode of operation will typically cause the concentrate grade to vary considerably throughout the production cycle in response to fluctuations in the feed characteristics. This problem may be overcome by feed homogenization to adsorb natural variations in feed quality throughout the production cycle. Alternatively, an automated control system can be implemented that blends feed streams from different quality stockpiles in proper portions just prior to separation (or blends concentrates just after separation).

Consider the simple case of a 500 tph preparation facility that operates only with heavy media circuits. The plant currently receives run-of-mine feed from two different coal seams. The primary seam, which is mined during three 8 hr shifts, is capable of providing a high yield at the target grade of 7.5% ash. In contrast, the second seam is very difficult to upgrade and, as such, is mined during only one 8 hr production shift. In order to select an appropriate control system for this plant, two sets of partition simulations were conducted over a 24 hr production period. The first set of computations was conducted to simulate the performance of a traditional control strategy. This scheme involved the realtime adjustment of specific gravity setpoints based on feedback from an online ash analyzer (Figure 6(a)). As discussed previously, this type of control system does not optimize yield. For comparison, another set of simulations were conducted in which the specific gravity setpoint was held constant and the feed blends were adjusted in response to feedback from the online analyzer to ensure that an acceptable product grade was maintained (Figure 6(b)). This approach attempts to maintain a constant incremental ash (and maximum yield) by holding the specific gravity setpoints at a constant value.

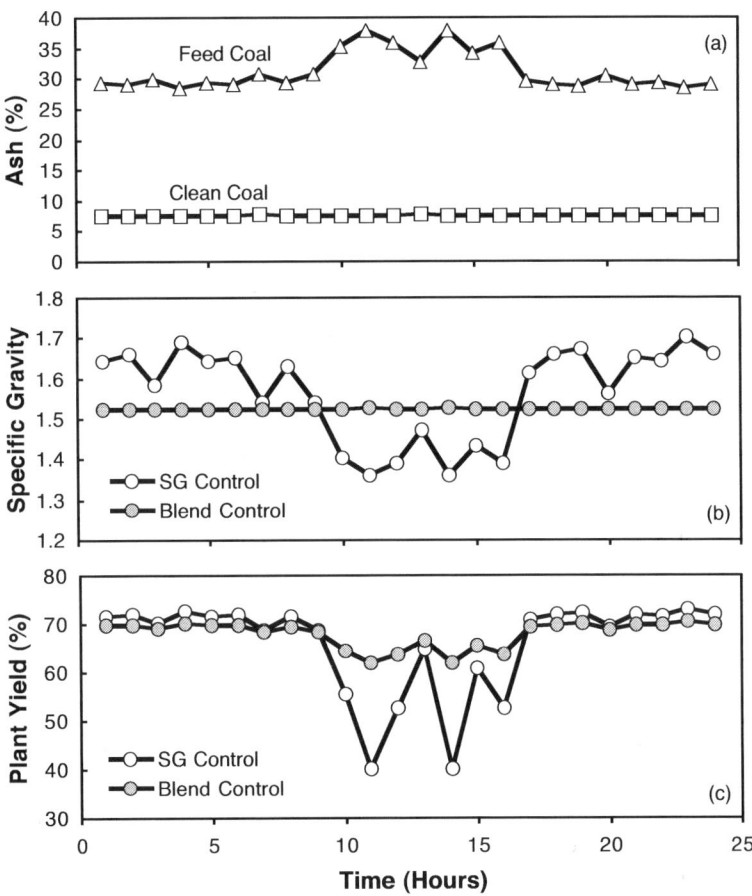

FIGURE 7 Comparison of plant performance based on the control of specific gravity cutpoints versus control of feed blend ratios for a 24 hr production cycle during which feed quality varied

Figure 7 compares the operating data from the two sets of control simulations. When the circuit is configured to automatically adjust the specific gravity setpoints, the ash content produced at any point in time remained relatively constant as a result of the online feedback from the analyzer (Figure 7(a)). However, this requirement forced the specific gravity setpoint to vary greatly over the production period (Figure 7(b)). This action was necessary to compensate for (i) natural variations in the quality of the two feed coals and (ii) the large disturbance created by the addition of the poor quality coal during the middle third of the 24 hr production period. In contrast, the control system based on feed blending provided a clean coal of consistent quality while keeping the same specific gravity setpoint (Figure 7(b)). In this case, the large surge of poor quality feed coal that would normally have to be dealt with during one shift is now spread out over the entire duration of the production period by the control system. The fluctuations in quality are handled by increases and decreases in the stockpile volumes of the two coals. More importantly, the blend control is based on the incremental grade concept that maximizes total plant yield. As a result, the blend based control system produced a total of 8,133 tons of clean coal over the 24 hr production period, compared to just 7,878 tons for the traditional control approach. This improvement provides 255 tons of additional saleable coal from the same tonnage of feed. As shown in Figure 7(c), the lower production associated with the traditional control strategy is due to sharp drops in yield during brief production periods. Obviously, it is not possible to make up this lost tonnage once it is discarded. In today's market, the higher production afforded by the improved control strategy represents a revenue increase of more than $1.6 million annually (i.e., 255 tons/day × $25/ton × 250 days/yr = $1,625,625). This clearly illustrates that the financial gains realized by proper implementation of existing processing engineering knowledge can be very substantial. Yet, despite these economic benefits, it is not uncommon to find plants in the eastern coalfields equipped with on-line ash analyzers that are used to control specific gravity cutpoints for grade control.

Analysis of Two-stage Spiral Configurations

Industrial situations often arise that can make use of both the incremental quality concept and circuit analysis. One such problem involves the design of spiral circuits for the coal industry. Spirals are one of the most effective methods for cleaning 1 × 0.15 mm coals. Unfortunately, the specific gravity cutpoints obtained using spirals are typically much higher than those employed by the coarse coal heavy medium circuits. The cutpoint difference becomes particular severe when plant capacity is pushed for economic reasons, yet no additional spirals are added to handle the increased tonnage. The large difference in cutpoint does not allow the overall plant to be optimized in accordance with the incremental quality concept. Also, water-based separators such as spirals tend to be inherently less efficient than heavy medium separators. For this reason, spirals are often used in two-stage circuits in which the middling streams from rougher spirals are rewashed using cleaner spirals. Plant operators must then discard the secondary middlings (and sacrifice yield) or retain the middlings (and accept a lower coal quality).

The process engineering concepts described earlier can be used to identify the best means for minimizing the inherent problems associated with coal spiral circuits. According to circuit analysis, the traditional rougher-cleaner spiral configuration has the same relative sharpness of separation as a single stage spiral. The optimal configuration would retreat the clean coal and middlings products from the rougher spirals in cleaner spirals. The cleaner spirals would be configured to recycle all material except the clean coal product back to the head of the circuit feed. This particular configuration provides a sharpness of separation that is 1.33 times higher than the traditional two-stage circuit. More importantly, this circuit provides a lower specific gravity cutpoint, which maximizes overall plant yield by better balancing the incremental quality between the spiral and heavy medium circuits. Unfortunately, this optimum spiral configuration is not practical due to the creation of large circulating loads. A more attractive configuration is to recycle only the middlings product from the cleaner spirals back to rougher feed. According to circuit analysis, this hybrid configuration improves the sharpness of the separation by 1.22 times compared to the traditional two-stage circuit. This circuit also has the advantage of reducing incremental ash since the retreatment of both the clean and middlings streams from the rougher spiral reduces the overall circuit cutpoint. While not as efficient as the optimum rougher-cleaner circuit, this configuration substantially reduces the amount of material that must be recycled and reduces the total number of spirals required.

Table 2 compares the performance of three different spiral configurations installed at the same industrial operation. These circuits include a traditional single-stage rougher circuit (Figure 8(a)), a rougher-cleaner circuit without middlings recycle (Figure 8(b)),

TABLE 2 Industrial comparison of different spiral circuit configurations

	Single-stage (No Recycle)	Two-stage (No Recycle)	Two-stage (Recycle)
Circuit Yield (%)	56.7	40.6	46.3
Circuit Ash (%)	18.8	9.05	9.23
Separation Sharpness (Ep)	0.18	0.20	0.15
Organic Efficiency (%)	90.4	83.4	94.3
SG Cutpoint	1.82	1.61	1.66

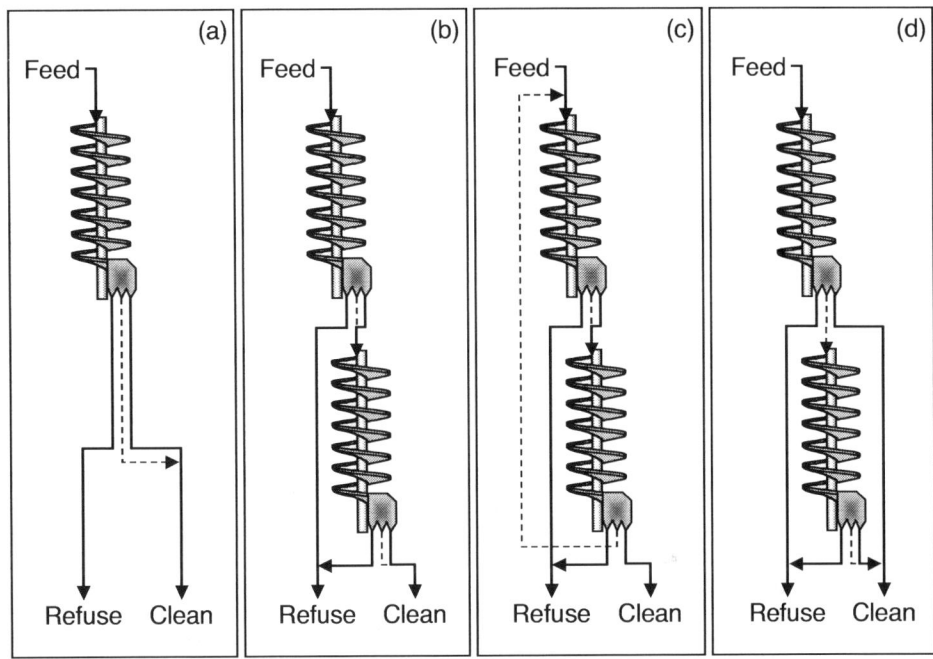

FIGURE 8 A comparison of several different spiral configurations: (a) single-stage rougher, (b) rougher-cleaner without recycle, (c) rougher-cleaner with middlings recycle, (d) traditional rougher with middlings cleaner

and a rougher-cleaner circuit with middlings recycle (Figure 8(c)). The traditional two-stage circuit that is commonly utilized in industry (Figure 8(d)) was not evaluated since this configuration is not capable of reducing the specific gravity cutpoint demanded by this exercise. The data showed that the single-stage spiral produced a clean coal yield of 56.7% and an unacceptably high ash of 18.8%. The poor ash was due to the high cutpoint (1.82 SG) and poor Ep (0.18). As expected, the rougher-cleaner circuit significantly lowered the cutpoint down to 1.61 SG and, in turn, reduced the clean coal yield and ash to 40.6% and 9.05%, respectively. However, this circuit reduced the organic efficiency from 90.4% to 83.4%. To correct this problem, circuit analysis indicates that the middlings stream must be recycled back to the head of the circuit. As shown in Table 2, this configuration significantly increased the clean coal yield to 46.3%, an improvement of 5.7 percentage points. This increase was achieved without substantially increasing the clean coal ash content (i.e., 9.23% vs. 9.05%). Also, the recycle circuit provided the best overall organic efficiency of 94.3%. It is also of interest to point out that the Ep for the rougher-cleaner circuit with recycle is 1.2 times lower than that of the single-stage circuit (i.e., 0.18/0.15 = 1.2). This ratio is in very good agreement with the theoretical value of 1.22 predicted by circuit analysis. For a three-shift operation with a circuit feed rate of about 45 tph, the improved circuit provides a revenue increase of more than $350,000 annually (i.e., 45 tph × 5.7% × $25/ton × 5,500 hr/yr = $352,700). Preliminary economic analyses show that this additional revenue offers an attractive payback on the capital investment. More importantly, the reduction in cutpoint from 1.82 to 1.66 SG allowed the incremental ash from the spiral circuit to be significantly improved. The lower ash allowed the specific gravity in the coarse coal heavy medium circuits to be raised. This modification, which is driven by the incremental quality concept, allowed nearly 13 tph of additional coal to be recovered from the coarse coal circuit. This incremental production provided additional revenue of nearly $1.8 million annually (13 tph × $25/ton × 5,500 hr/yr = $1.79 MM). Once again, the application of processing engineering knowledge allowed plant profitability to be significantly improved.

CONCLUSIONS

Process engineering principles, such as the incremental quality concept and circuit analysis, provide extremely useful information regarding the design, optimization and control of density-based separation processes. Unfortunately, field personnel and plant designers often seem to be unaware of these powerful tools or fail to recognize how they may be applied to solve industrial problems. In light of this shortcoming, several practical examples have been provided that show the potential financial gains that may be realized by implementing these versatile principles. These wide-ranging examples include the optimization of heavy medium circuit performance, the design of two-stage spiral circuits, and the realtime control of plant grade. In each case, the application of processing engineering knowledge helped to identify the most appropriate solution to each industrial problem. Personnel working the mineral and coal processing industries are encouraged to learn more about these versatile concepts and the benefits they offer. Educators and researchers can also serve an important role by promoting the use of processing engineering principles to their industrial counterparts.

ACKNOWLEDGMENTS

The author would like to acknowledge the contributions that have been made to this work by industrial parties working in the coal and mineral processing industries. Particular gratitude is expressed to Fred Stanley at the Pittston Company and Peter Bethell at Massey Services. The funding provided by the U.S. Department of Energy (DOE) in support of various field projects that relate to this topic is also gratefully acknowledged.

REFERENCES

Abbott, J., 1982. The Optimisation of Process Parameters to Maximise the Profitability from a Three-Component Blend, 1st Australian Coal Preparation Conf., April 6–10, Newcastle, Australia, 87–105.

Anonymous, 1966. Plotting Instantaneous Ash Versus Density, *Coal Preparation*, Jan.–Feb., Vol. 2, No. 1, p. 35.

Armstrong, M. and Whitmore, R.L, 1982. The Mathematical Modeling of Coal Washability, 1st Australia Coal Preparation Conf., April 6–10, Newcastle, Australia, 220–239.

Dell, C.C., 1956. The Mayer Curve, *Colliery Guardian*, Vol. 33, pp. 412–414.

Luttrell, G.H., Barbee, C.J., and Stanley, F.L., 2003. "Optimum Cutpoints for Heavy Medium Separations," *Advances in Gravity Concentration*, Symposium Proceedings, Society for Mining, Metallurgy, and Exploration, Inc., (SME), Littleton, Colorado.

Lyman, G.J., 1993. Computational Procedures in Optimization of Beneficiation Circuits Based on Incremental Grade or Ash Content, *Trans. Inst. Mining and Metallurgy*, Section C, 102: C159–C162.

Mayer, F.W., 1950. A New Washing Curve. *Gluckauf*, Vol. 86, pp. 498–509.

Meloy, T.P., 1983. "Analysis and Optimization of Mineral Processing and Coal-Cleaning Circuits—Circuit Analysis," *International Journal of Mineral Processing*, Vol. 10, p. 61.

Rong, R.X. and Lyman, G.J., 1985. Computational Techniques for Coal Washery Optimization—Parallel Gravity and Flotation Separation, *Coal Preparation*, 2: 51–67.

SECTION 1

Gravity Concentration Fundamentals

- On the Phenomena of Hindered Settling in Liquid Fluidized Beds **19**
- Modeling of Hindered-settling Column Separations **39**
- Dense Medium Rheology and Its Effect on Dense Medium Separation **55**
- Methodology for Performance Characterization of Gravity Concentrators **71**

On the Phenomena of Hindered Settling in Liquid Fluidized Beds

K.P. Galvin[*]

Liquid fluidized beds are being used increasingly in gravity separation and hydrosizing of mineral particles. In these applications, amongst others, it is essential to understand the hindered settling of suspensions that consist of a broad range of particle sizes and density. Fundamentally, these suspensions are exceedingly complex and hence, despite their practical importance, are invariably neglected when hindered settling models are investigated. Unfortunately the academic focus tends to be on "simple" binary suspensions. In this paper, existing hindered settling models are reviewed in the context of phenomena that occur in fluidized beds, such as inversion. Batch fluidization methods for obtaining fundamental information on gravity separation, such as grade recovery curves, partition curves, and slip velocity data are described. New developments in liquid fluidized beds, utilizing sets of parallel inclined plates to increase segregation rates, are also discussed.

INTRODUCTION

Hindered settling is the term used to describe the influence of neighboring particles on the settling velocity of a given particle species in a suspension. Clearly, an understanding of the phenomena of hindered settling is fundamental to describing the separation processes that occur in gravity separation and size classification devices. With such an understanding, it should be possible to develop accurate process models that are applicable to a broad range of conditions, and in turn optimize process flow-sheets or even develop new or improved separation processes.

This paper is concerned with the phenomena of hindered settling in fluidized beds. These systems operate under uniform flow conditions, and thus provide well-defined conditions for investigating the factors that govern hindered settling. There is also growing interest in using fluidized bed devices to achieve gravity separation and also size classification, and hence there are direct benefits to be gained through improving our understanding of fluidized beds.

---

[*] School of Engineering, University of New Castle, Callaghan, NSW, Australia.

In mineral processing we need a description of hindered settling that is applicable to suspensions that consist of a broad range of particle sizes and densities. In the past, rigorous experimental validation of a hindered settling model has only involved simple suspensions consisting of typically mono or binary species (Lockett and Al-Habbooby, 1973, 1974; Al-Naafa and Selim, 1992; Davis and Gecol, 1994). The vast majority of studies involved particles of just one density and, in general, a single size species was actually approximated by a "narrow" distribution of particle sizes. This limited form of validation has achieved varying levels of success, dependent upon the specific system of particles. It is questionable, therefore, whether there exists a satisfactory description of hindered settling that is reliable for general use in mineral processing.

This paper consists of three parts. The first is a review of hindered settling in fluidized beds. The phenomenon of phase inversion in fluidized beds, which is central to understanding how a fluidized bed can be used to achieve either a gravity separation or a size classification, is discussed. The phenomenon has led to some debate on whether the buoyancy force on a particle depends on the density of the liquid or the density of the suspension. This issue is central to the validity of using macroscopic suspension properties such as density and viscosity to account for the effects of hindered settling. Three classes of models for describing the slip velocity and hindered settling of a particle are then introduced and examined in more detail.

The second part of the paper is concerned with a series of related studies involving fluidized beds. The focus is on feeds of coal and mineral matter. A new fluidization method (Callen et al., 2002a, 2002b) for obtaining a detailed description of the size and density distribution of such feeds is described. This method has the potential to replace the sink-float method in coal preparation, and to provide a method for analyzing minerals of much higher density. Further analysis of the experimental data is then used to obtain slip velocity data suitable for quantifying the separation produced in a fluidized bed. Those results are then compared with values predicted using a generalized slip velocity model.

In the third part of the paper, we examine the effect of introducing sets of parallel inclined plates into a fluidized bed, and comment on the range of benefits achieved.

PART I—PHENOMENA IN FLUIDIZED BEDS

In this section the subject of hindered settling is reviewed with reference to the phenomenon of phase inversion in fluidized beds. Then classes of slip velocity models that have been proposed to account for hindered settling are discussed.

We begin by examining the concept of hindered settling. Consider a spherical particle of diameter, d, density ρ_p, and volume v, settling in a liquid of density, ρ, and viscosity, μ. The weight of the particle, $\rho_p g v$, is balanced by the sum of the drag force, F_d, and the buoyancy force, $\rho g v$. Hence, the drag force can be equated with the net weight of the particle in the fluid. That is,

$$F_d = (\rho_p - \rho) g v \qquad \text{(EQ 1)}$$

For an isolated spherical particle settling at its terminal velocity, U_t, with a low Reynolds number less than about 0.1, the drag force is given by the Stokes law, $F_d = 3\pi\mu U_t d$. Setting $v = (\pi/6)d^3$, the terminal velocity is,

$$U_t = \frac{(\rho_p - \rho) g d^2}{18\mu} \qquad \text{(EQ 2)}$$

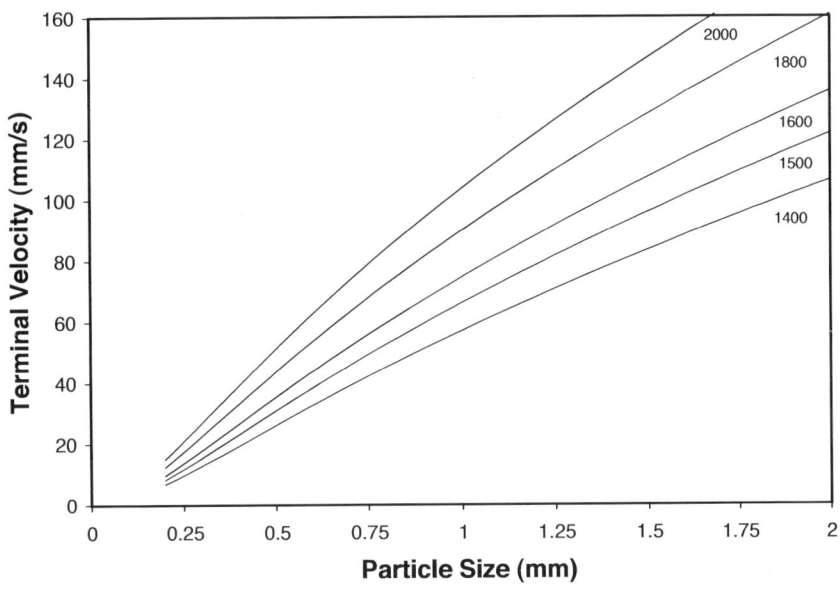

FIGURE 1 Particle terminal velocities in water based on the equation of Zigrang and Sylvester (1981). The label on each curve denotes the particle density in kg/m³.

At higher Reynolds numbers, Re, the terminal velocity of a particle can be obtained using the following empirical equation of Zigrang and Sylvester (1981). That is,

$$Re = \frac{\rho U_t d}{\mu} = \left[\left(\frac{14.51 + (g(\rho_S - \rho)\rho)^{0.5} 1.83 d^{1.5}}{\mu}\right)^{0.5} - 3.81\right]^2 \quad \text{(EQ 3)}$$

Results relevant to this study, calculated using Equation 3, are shown in Figure 1.

When the test particle resides in a suspension, it is appropriate to refer to the slip velocity, U_{slip}, which is defined as the velocity of the particle, U_s, relative to the interstitial velocity of the liquid, U_f. That is,

$$U_{slip} = U_s - U_f \quad \text{(EQ 4)}$$

A test particle in a suspension experiences a resistance to motion that is dependent upon the neighbouring state of the suspension. At low Reynolds numbers, the drag force acting on the test particle is directly proportional to the slip velocity. That is,

$$Fd = \frac{3\pi \mu d U_{slip}}{F} \quad \text{(EQ 5)}$$

where F, the hindered settling factor, is defined as the ratio of the resistance to motion experienced by the test particle in isolation, relative to that in the suspension. The slip velocity of the test particle is reached when the drag force equals the net weight of the test particle in the liquid, as defined by Equation 1. Thus, by equating Equations 1 and 5, and inserting Equation 2,

$$F = \frac{U_{slip}}{U_t} \quad \text{(EQ 6)}$$

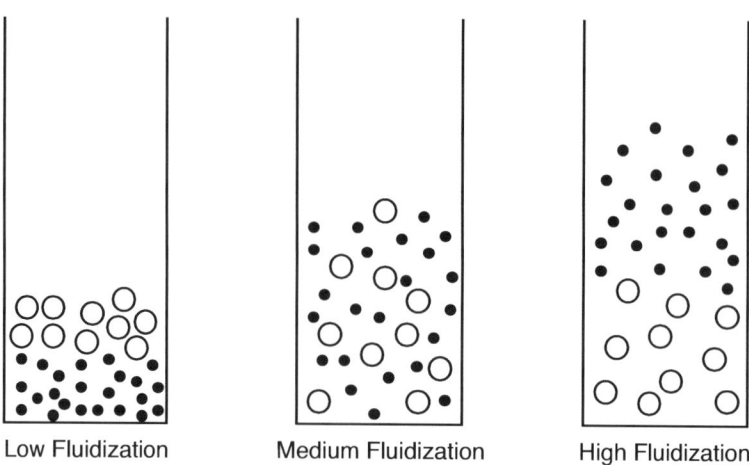

FIGURE 2 Representation of inversion in a fluidized bed using a binary system of large low density particles (open circles) and small high density particles (filled circles). From left to right the fluidization rate increases from low to medium to high.

and hence the hindered settling factor is the ratio of the two velocities, the slip velocity of the particle species, U_{slip}, in the suspension and the terminal velocity, U_t, of the same particle species in isolation.

Equation 6 is well illustrated by considering a liquid fluidized bed containing identical spherical particles. The relationship between the superficial velocity, U, and the volume fraction of the solids, ϕ, is described empirically using the equation of Richardson and Zaki (1954). That is,

$$U = U_t(1-\phi)^n \quad \text{(EQ 7)}$$

where n is an exponent. Here, the particle velocity relative to the vessel is zero, and hence the slip velocity of the particles is,

$$U_{slip} = \frac{U}{(1-\phi)} = U_t(1-\phi)^{n-1} \quad \text{(EQ 8)}$$

It follows from Equation 6 that the hindered settling factor is,

$$F = (1-\phi)^{n-1} \quad \text{(EQ 9)}$$

Equation 9 has also been applied to suspensions consisting of more than one species, with ϕ representing the total volume fraction of all the species (Lockett and Al-Habbooby, 1973; 1974).

Inversion in Fluidized Beds

The phenomenon of inversion in liquid fluidized beds (Moritomi et al., 1982), shown in Figure 2, illustrates the serious limitation of many slip velocity models, especially those dependent on the volume fraction of the solids only (Asif, 1998). Figure 2 shows the fluidization of a binary suspension of relatively fine dense particles and larger less dense particles at a low, medium, and high rate of fluidization. In isolation, the larger, less dense, species has the higher terminal velocity. At a low superficial liquid velocity a high suspension density is produced, with the finer species resident below the coarser species. However, as the fluidization rate is increased, and in turn the suspension density lowered, the two species mix. This mixing could, of course, be a consequence of dispersion, brought about by the increase in the fluidization rate. However, at a higher fluidization rate the mixing actually disappears and the system finally inverts. It is reasonable to conclude, therefore, that the intermediate mixing state does not arise from a mechanism of dispersion. Rather, the two species adopt a common slip velocity.

Imagine that at the low fluidization rate the two particle species were artificially mixed to produce a combined volume fraction of solids, ϕ. When the mixing ceases, the finer particles would move downwards and the coarser particles upwards relative to the vessel. Hence, the slip velocity of the finer particles is greater than for the coarser particles. When in the mixed state, the slip velocity of each species is the same, and after inversion the slip velocity of the coarser species is higher than for the finer species. The Richardson and Zaki model, given by Equation 9, and other models based on the volume fraction only, predict a common hindered settling factor for both species, and hence fail to predict the inversion phenomenon.

It is evident from the inversion phenomenon that gravity separation is promoted by using relatively low fluidization rates and hence high suspension densities. By increasing the fluidization rate, which in turn lowers the suspension density, a separation more dependent on the particle size is achieved, though particle density will continue to play a role. These observations have led to a debate concerning the buoyancy force on a particle in a suspension, whether the force depends upon the density of the suspension or the density of the liquid (Clift et al., 1987).

Slip Velocity Models

There are three basic classes of models for describing the slip velocity and hence the hindered settling factor of a particle species. The first, and the most common, are models that describe the free settling of a given species in a liquid while in the presence of other particles. Usually these models are expressed in terms of the volume fraction of the solids only, and hence cannot explain the phenomenon of inversion. Examples include the so-called cell models (Di Felice (1995)), derived from the Navier Stokes equations, as well as the model by Batchelor (1972), based on a random distribution of particles around a test sphere. Batchelor (1982) and Batchelor and Wen (1982) extended this work to cover multi-species systems consisting of particles having different sizes and densities at low volume fractions. Brauer and co-workers (1966, 1973) developed a novel and complex model that took into account the effects of turbulence.

Despite the obvious appeal of the more fundamental models, empirical equations continue to be widely used. The best known is the Richardson and Zaki (1954) equation, defined by Equation 9. Lockett and Al-Habbooby (1973, 1974) extended this equation to cover more than one particle species, with ϕ denoting the sum of the volume fractions of all the species. At low Reynolds numbers, the value of the exponent, n, was originally reported to be 4.65. Garside and Al-Dibouni (1977) suggested n = 5.1 for low Reynolds numbers, effectively arriving at the general result,

$$n = \frac{5.1 + 0.27 \text{Re}^{0.9}}{1.0 + 0.1 \text{Re}^{0.9}} \qquad \text{(EQ 10)}$$

The theory of Batchelor (1972), which in effect predicts the value to be as high as 6.55, is well supported by the definitive work of Al-Naafa and Selim (1992).

The second class of model is described by Zuber (1964), and discussed further in the book by Brodkey (1967). The hindered settling is accounted for by firstly assuming the particle settles in a medium having an effective viscosity, μ_s. Here, the particle is assumed to experience an additional force based on the dissipative pressure gradient, dP/dZ, of the surrounding suspension rather than the total pressure gradient as used by Brodkey (1967). The steady state force balance becomes,

$$\rho_p g v = F'_d + \frac{v dP}{dZ} + \rho g v \qquad \text{(EQ 11)}$$

where the modified drag force is given by,

$$F'_d = 3\pi \mu_s U_{slip} d \qquad \text{(EQ 12)}$$

at low Reynolds numbers. The dissipative pressure gradient is given by,

$$\frac{dP}{dZ} = (\rho_s - \rho)\phi g \qquad \text{(EQ 13)}$$

where ρ_s is the average density of the solids. If the average density of the solids is identical to the density of the test particle, substitution of the modified drag force and the dissipative pressure gradient into Equation 11, and then comparing the result with Equations 2 and 8, gives

$$\frac{\mu_s}{\mu} = (1-\phi)^{2-n} \qquad \text{(EQ 14)}$$

Einstein (1906, 1911) developed an equation for the effective viscosity of suspensions at exceedingly low volume fractions. That is,

$$\frac{\mu_s}{\mu} = 1 + 2.5\phi \qquad \text{(EQ 15)}$$

Brinkman (1952) used a recursive approach to extend the application of Einstein's equation to cover much higher volume fractions. This leads to the differential equation,

$$\frac{d\mu_s}{d\phi} = \frac{2.5\mu_s}{1-\phi} \qquad \text{(EQ 16)}$$

Integration gives the extended result,

$$\frac{\mu_s}{\mu} = (1-\phi)^{-2.5} \qquad \text{(EQ 17)}$$

in reasonable agreement with the higher order dependence reported by Batchelor and Green (1972). Other relationships for suspension viscosity are provided by Barnea and Mizrahi (1973).

Equations 14 and 17 suggest n = 4.5, and hence that the Richardson and Zaki equation can be arrived at by using an effective viscosity and density based on the macroscopic suspension properties. Given the excellent agreement between the theory of Batchelor (1972) and the experimental results of Al-Naafa and Selim (1992), with n effectively equal to 6.55, one should proceed immediately with caution. The present discussion concerns a test particle identical to those that form the suspension. In principle, the particles surrounding our test particle exhibit a zero velocity relative to our particle, and hence do not participate in the deformation of the flow around the particle, though they do generate higher velocity gradients and hence stresses in the liquid (Zuber, 1964), where all energy dissipation occurs (Clift et al., 1987). The validity of equation 14 is questionable. It should be noted, however, that a test particle placed in a suspension of very much finer particles does experience the suspension viscosity and density (Batchelor, 1982).

Masliyah (1979) produced a more generalized model suitable for describing the slip velocity of a particle settling in a suspension of particles either of the same or of different density. Expressing the dissipative pressure gradient in terms of the suspension density, ρ_{susp}, gives,

$$\frac{dP}{dZ} = (\rho_{susp} - \rho)g \qquad \text{(EQ 18)}$$

Combining this result with Equations 2, 6, 11, and 12, and using Equation 14 to describe the suspension viscosity, leads to,

$$F = \frac{(1-\phi)^{n-2}(\rho_p - \rho_{susp})}{(\rho_p - \rho)} \qquad \text{(EQ 19)}$$

The first term on the right hand side accounts for the effect of the suspension viscosity and the second for the effect of the suspension density. Again, despite the physical appeal of this result, the model must be regarded as entirely empirical. Asif (1997) and later Galvin et al. (1999) strengthened the empirical dependence on the suspension density, arriving at,

$$F = \left(\frac{\rho_p - \rho_{susp}}{\rho_p - \rho}\right)^{n-1} \qquad \text{(EQ 20)}$$

Equations 19 and 20 revert to the Richardson and Zaki model when all particles are of the same density. However, for suspensions consisting of particles of different density, these equations offer some explanation for inversion in fluidized beds. Galvin and co-workers (1999a, 1999b, 2001) have had some success in using Equation 20 to describe fluidized bed systems, but no rigorous validation over a broad range of conditions has been undertaken. It should be noted that Equation 20 is undefined for circumstances involving a particle of density lower than the suspension density, however, such suspensions are invariably non-homogeneous.

The third class of model is applicable to very high volume fractions, bordering on the conditions of a packed bed. A weakness of this approach is its failure to adequately describe the effects of the fluid drag in the dilute limit. Foscolo et al. (1983) and later Gibilaro et al. (1986) addressed this problem using a modified packed bed model to describe the inversion behaviour observed in fluidized beds. The approach attracted criticism (Clift et al., 1987), the central concern again being whether the buoyancy force should be based on the density of the continuous liquid or on the density of the suspension. Nevertheless, the approach has proved successful in describing the fluidized bed inversion phenomenon.

PART II—FLUIDIZATION STUDIES

In this section, some recent fluidization work concerned with generating data on the size and density distribution of a particle feed is described. Knowledge of the size and density distributions is fundamental to describing the gravity separation and size classification of actual feeds consisting of a broad range of particle sizes and densities, including the grade recovery curve. The measurement of the density distribution is far from trivial and poses increasing concerns on health and environmental grounds when dense liquids need to be used. It is also shown in this section that the fluidized bed method can be extended to obtain experimental measurement of particle slip velocity data that is directly relevant to the gravity separation of the feed. These data are compared with the predictions from Equation 20.

Measurement of the Size and Density Distribution of a Feed by Fluidization

Fluidized beds can be used to quantify the size and the density distribution of a given feed. Indeed, Galvin and Pratten (1999) have suggested that water fluidization could replace the sink-float method used in coal preparation, thus overcoming the health and environmental problems associated with using heavy liquids. Callen et al. (2002a, 2002b) have since conducted a much more detailed investigation of this approach using a range of coal feeds. Two fluidization methods were examined and validated against the sink-float method by examining the cumulative yield as a function of the particle density and also the particle ash.

In the first method, the feed was initially dry sieved into a series of narrow size fractions using the following screens: 0.5, 0.7, 1.0, 1.4, and 2.0 mm. Each size fraction was then independently fluidized, with the lowest density particles migrating to the top of the bed, leaving increasingly denser particles below. Once an equilibrium state was reached, the suspension was sampled through valves in the side of the vessel as shown in Figure 3. All of the suspension located above the highest valve was discharged through the highest valve. The second highest valve was then opened and all of the suspension above this valve discharged etc. In this way a series of samples corresponding to different elevations of the vessel were produced. The average density of the particles present in each sample was then measured using pycnometry, and the ash content determined.

In the second method a single fluidization was conducted using the overall sample, 0.5–2.0 mm. A relatively high fluidization rate was used initially to ensure all of the particles were suspended. The rate was also oscillated for a brief period and then lowered to a fixed rate. At the lower fluidization rate, a high proportion of the particles formed a packed bed on the base of the vessel. The remaining particles, however, were fully suspended and distributed, with the slowest settling particles near the top. Samples of the suspended particles were withdrawn from the top down to a level just above the packed bed. The fluidization rate was then increased in order to suspend a portion of the packed bed. Further samples of suspended particles were then removed as before. This cycle was repeated until the contents of the vessel were fully discharged. Each of the samples removed from the vessel were then sieved using the same set of screens as before: 0.5, 0.7, 1.0, 1.4, and 2.0 mm, and each of the size fractions analysed to determine the particle density and ash content.

Results showing the cumulative yield versus particle density and the cumulative yield versus cumulative ash are shown in Figures 4A and 4B respectively. In these figures the float sink method is compared with the results obtained using methods one and two. For the purpose of brevity, the composite result of all of the size fractions is presented here,

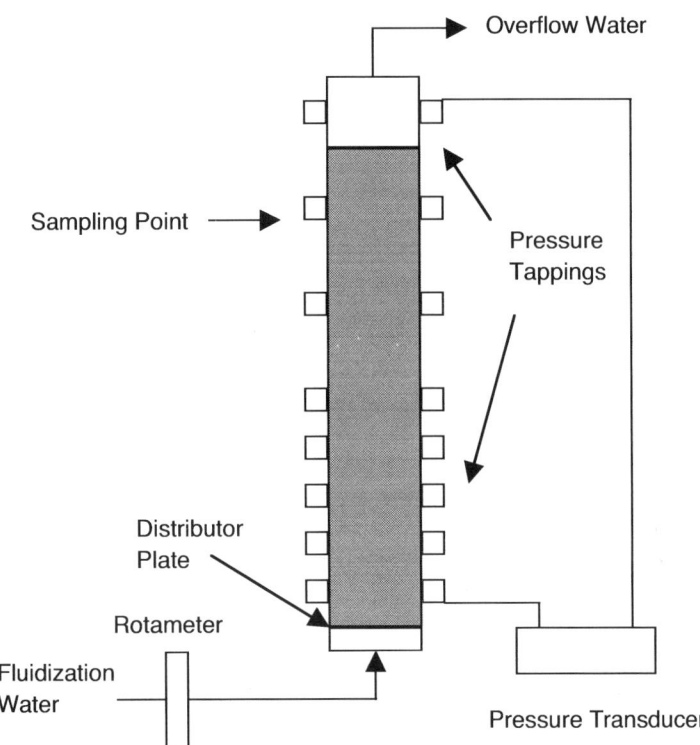

FIGURE 3 Schematic representation of the fluidization vessel showing sampling locations and tappings for the pressure transducer

rather than separate results for each of the narrow size fractions. It is evident that there is excellent agreement between the results from the two fluidization methods in both Figures 4A and 4B indicating that the final result is independent of whether the particles are sized and then fluidized, or fluidized and then sized. Further, the cumulative yield versus density plot indicates excellent agreement between the sink-float method and the two fluidization methods. There is, however, an obvious discrepancy between the cumulative yield versus cumulative ash results, with the sink-float method producing the higher yield at a given ash.

The discrepancy between the results obtained by the sink-float and fluidization methods is due to an inherent physical phenomenon in fluidized beds known as dispersion. Dispersion, which is a form of mixing, is driven by concentration gradients in a manner analogous to molecular diffusion. The dispersion coefficient, however, can vary significantly, and will tend to increase as the fluidization rate increases (Asif and Peterson, 1993). Callen et al. (2002c) examined the dispersion of a binary system of particles, with the density of one species 1,300 kg/m^3 and the other 1,400 kg/m^3. The two species were from the same narrow size range of 1.18 to 1.40 mm. Ramirez and Galvin (2002) developed a continuum model based on the approach of Kennedy and Bretton (1966) to describe the processes of dispersion and sedimentation in a fluidized bed containing a binary system of particles under batch conditions. Their model was validated using the data of Callen et al. (2002c). The governing equation for a single species i is given by the net particle flux relative to the vessel. That is,

28 | GRAVITY CONCENTRATION FUNDAMENTALS

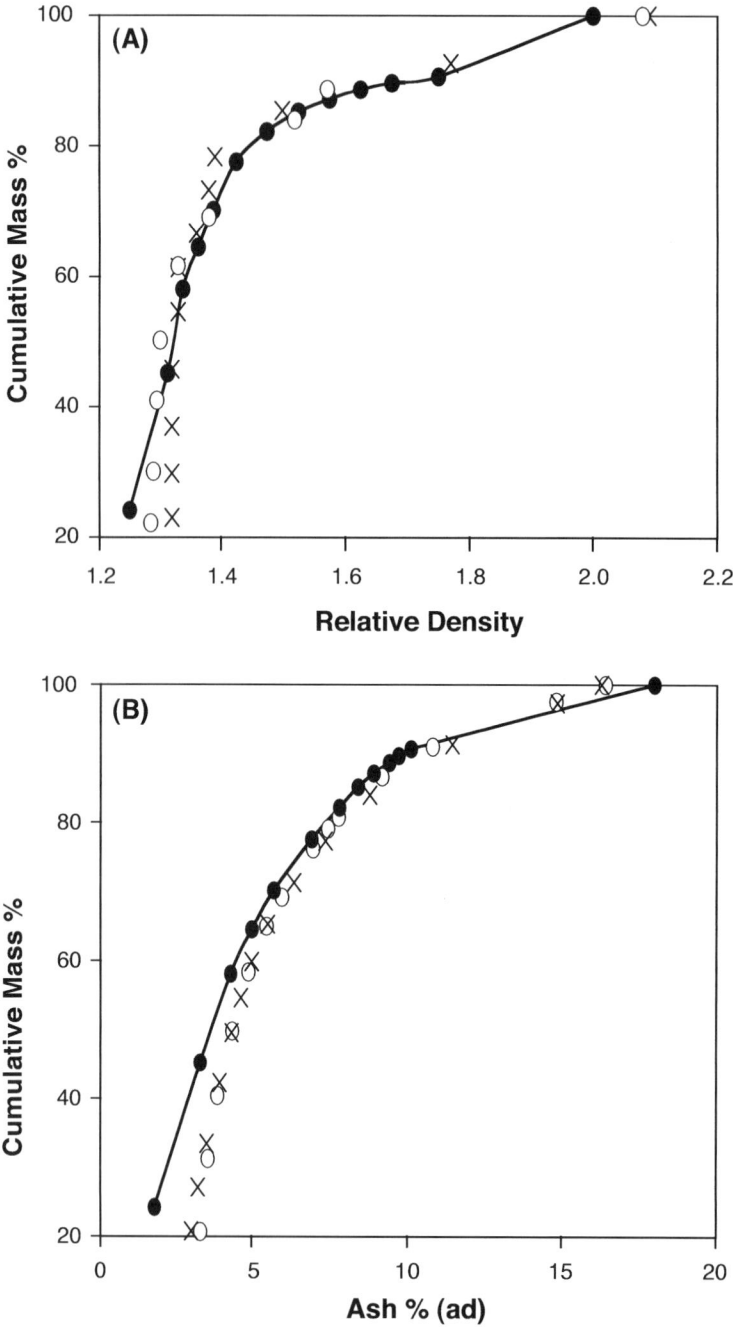

FIGURE 4 (A) Cumulative yield versus relative density, and (B) cumulative yield versus ash curves for the −2.0 +0.5 mm size fraction. The dark circles represent the data generated by the traditional float-sink method. The crosses and open circles represent the results for the first and second methods respectively.

$$N_i = U_{pi}\phi_i = \frac{-D_i \partial \phi_i}{\partial z} + U_{si}\phi_i \qquad (EQ\ 21)$$

where U_{pi} is the velocity of species i relative to the vessel in the presence of dispersion, and U_{si} is the velocity of species i relative to the vessel in the absence of dispersion. In order to obtain the equilibrium solution, they solved the dynamic flux balance,

$$\frac{\partial \phi_i}{\partial t} = \frac{-\partial N_i}{\partial z} \qquad (EQ\ 22)$$

for each species. At equilibrium the net flux, N_i, of each species relative to the vessel is zero. There is then a steady state balance between the dispersive flux, $-D_i \partial \phi_i / \partial z$, and the sedimentation flux, $U_{si}\phi_i$. The interstitial fluid velocity is $U/(1-\phi)$. Although the particles exhibit no net movement relative to the vessel, they do exhibit a sedimentation velocity relative to the vessel, which can only be calculated using the dispersive flux. Ideally, in obtaining experimental slip velocity data the potential for dispersion to impact on the data obtained should be recognised and, if not accounted for, should be minimized by lowering the concentration gradients produced by a feed using a relatively tall fluidized bed.

Measurement of Slip Velocities

In this section we examine in detail data obtained using the second fluidization method, the aim being to obtain information on the particle slip velocities. As already noted, the fluidization rate is raised and even oscillated in order to permit particles incorrectly positioned in the bed at the base of the vessel to be released and in turn suspended. The fluidization rate is then lowered, and the system allowed to reach equilibrium at a fixed superficial velocity. Thus the final state consists of a packed bed of relatively fast settling particles at the base, and a suspension above. In the experiment, the suspended particles undergo a self-sorting process, with those particles capable of adopting identical slip velocities co-locating at the one elevation.

A matrix of data obtained using the second fluidization method is shown in Tables 1–3. Each row of data represents a sample from the fluidization vessel, with the top row corresponding to the uppermost sample etc. A standard size analysis of the particles present in each of the samples, obtained using the usual sequence of sieves referred to above, is shown in Table 1. The average density of the particles present in each of the size fractions, determined by pycnometry, is shown in Table 2. These data provide a relationship between the particle size and particle density for particles that settle at a common slip velocity. For a superficial fluidization velocity of U and suspension containing particles with a total volume fraction of ϕ, the particles forming the suspension have a common slip velocity, U_{slip}, of $U/(1-\phi)$, assuming the effects of dispersion can be neglected. These results are listed in Table 3. As shown in this table, two main sets of data were generated using the superficial fluidization velocities, 13.9 and 21.0 mm/s. The higher rate allowed faster settling particles, previously comprising the bed on the base of the container, to participate in the experiment by being suspended.

Figure 5 shows a series of curves describing combinations of particle size and density that yield a common slip velocity. Each curve was generated using Equations 3, 10, and 20, based on the suspension density and slip velocity values shown adjacent to the curve, and listed in Table 3. The suspension densities here are relatively low and hence Equations 19 and 20 lead to similar results. It is noted that sample 1 was neglected because only two size fractions were represented, and that sample 8 was not considered

TABLE 1 Distribution showing the mass % at a given level in the vessel falling within a specified size range. The total mass of solids was 0.991 kg.

Mass (%)	Size Fraction (mm)				
Sample	−2.0+1.4	−1.4+1.0	−1.0+0.71	−0.71+0.50	−2.0+0.50
Mean Size (mm)	1.700	1.200	0.855	0.605	
1	0.0	0.0	1.2	5.0	6.2
2	0.0	0.7	6.1	3.8	10.6
3	0.0	5.8	5.4	2.2	13.4
4	3.3	8.6	2.9	1.3	16.1
5	8.3	5.2	1.5	0.9	15.9
6	7.3	3.1	1.1	0.9	12.4
7	7.5	2.8	0.9	1.2	12.4
8	7.0	3.8	1.1	1.1	13.0
Total	33.4	30.0	20.2	16.4	100.0

TABLE 2 Distribution showing the relative density of the particles at a given level in the vessel falling within a specified size range.

Relative Density	Size Fraction (mm)				
Sample	−2.0+1.4	−1.4+1.0	−1.0+0.71	−0.71+0.50	−2.0+0.50
Mean Size (mm)	1.700	1.200	0.855	0.605	
1			1.26	1.32	1.31
2		1.29	1.29	1.36	1.32
3		1.29	1.32	1.41	1.32
4	1.27	1.30	1.34	1.50	1.32
5	1.30	1.33	1.39	1.55	1.33
6	1.31	1.37	1.43	1.74	1.36
7	1.37	1.46	1.51	1.90	1.44
8	1.73	1.96	2.11	2.23	1.86

TABLE 3 Information on the height of the middle position of each sample, the superficial fluidization velocity, suspension density, average solids density, volume fraction of solids, and slip velocity. Note that the superficial fluidization velocity was increased in order to cause the suspension of the lower potion of the system, producing a reduction in the suspension density and volume fraction of solids. Note also that the particles on the base of the container were not fully fluidized, as evident by the relatively low suspension density (measured by a pressure transducer) and volume fraction of solids, and relatively high particle density.

Sample	Height (m)	U (mm/s)	ρ_{susp} (kg/m³)	ρ_s (kg/m³)	ϕ	U_{slip} (mm/s)
1	1.365	13.9	1,037	1309	0.120	15.8
2	1.165	13.9	1,054	1317	0.169	16.7
3	0.965	13.9	1,074	1319	0.232	18.1
4	0.765	13.9	1,092	1316	0.291	19.6
5	0.565	13.9	1,112	1328	0.342	21.1
6	0.380	21.0	1,100	1360	0.278	29.1
7	0.170	21.0	1,142	1440	0.325	31.1
8	0.068	21.0	1,266	1860	0.309	30.4

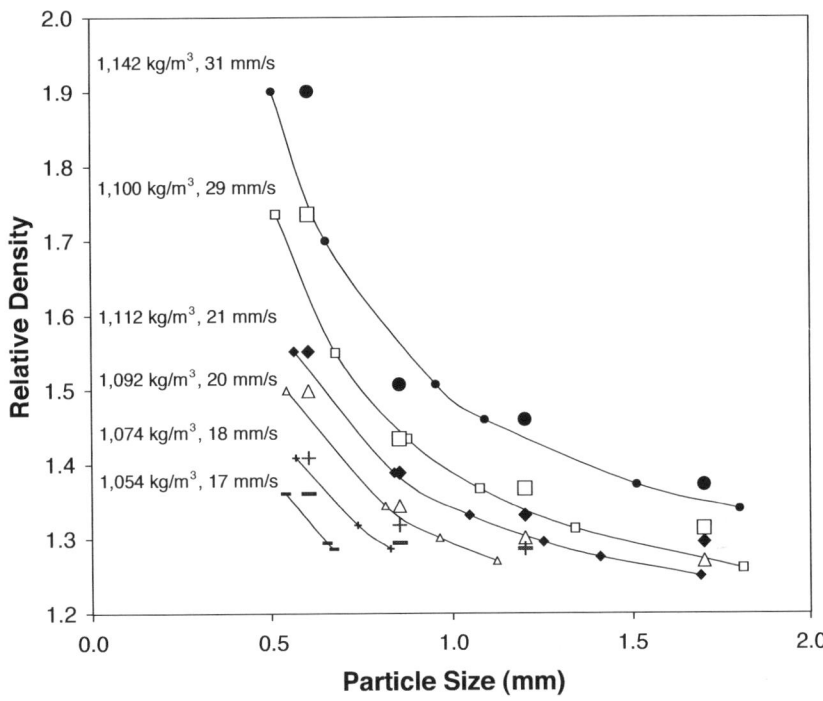

FIGURE 5 The combination of particle species having the same slip velocity are represented by a plot of the species relative density versus its particle size. For example, a small dense particle will tend to settle at the same rate as a large low density particle. The curves are the model predictions based on the specified suspension density and slip velocity values adjacent to the curves. The curves are identified by the small symbols and the experimental data by the corresponding large symbols. The model results compare well with the experimental data for the same conditions. The experimental data are listed in Table 3.

because the particles were not fully suspended. Each curve is defined by a set of small symbols, and should be compared with the experimental data denoted by the corresponding set of large symbols. In general, the agreement is remarkably good. However, this limited validation does not ensure the model will work under all conditions, especially with particles of much higher density.

A feed that consists of a special combination of particles having a common slip velocity will tend to report equally to the overflow and underflow. Other particle species, introduced to the feed, will exhibit either a lower or higher slip velocity, and hence report to the overflow or the underflow respectively. Thus, the special combination of particles behave as D_{50} particles, with an equal likelihood of partitioning to either of the output streams. This concept is used later in the paper to describe the separation achieved in a novel fluidized bed separator.

PART III—NEW DEVELOPMENTS IN FLUIDIZED BEDS

In a conventional fluidized bed consisting of identical particles, there is a unique relationship between the superficial fluidization velocity and the volume fraction of the solids. Once fluidized, the bed expands according to the empirical equation of Richardson

32 | GRAVITY CONCENTRATION FUNDAMENTALS

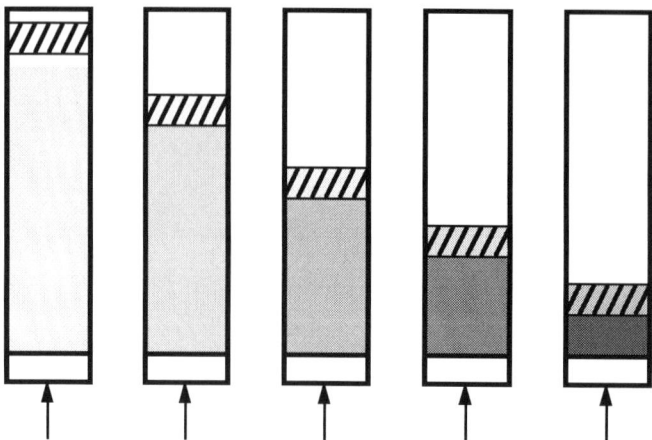

Fluidization water up through vessel

FIGURE 6 A conventional fluidized bed containing a set of parallel inclined plates. The set of plates is gradually moved downwards, with the fluidization rate held constant. The suspension is retained below the set of plates, producing a significant concentration increase. Clearly, the system concentration is no longer directly dependent on the fluidization rate only.

and Zaki (1954). Galvin and Nguyentranlam (2002) introduced a set of parallel inclined plates into a fluidized bed as shown in Figure 6. These plates produced a series of inclined channels, and hence a significant segregation effect. The fluidized suspension flowed up through the inclined channels, with particles depositing onto the incline, forming sediment, and then sliding back down to the suspension below. The high segregation rate is equivalent to the so-called Boycott Effect (Boycott, 1920) that forms the basis of the lamellae thickener. The increased segregation produced through this arrangement results in the retention of a suspension below the top of the inclined plates, even when the fluidization rate exceeds the terminal velocity of the particle species. In particular, there is no longer a one-to-one relationship between the superficial fluidization velocity and the volume fraction of solids. A given fluidization rate can support a broad range of volume fractions.

Galvin and co-workers (2001, 2002a) have conducted a number of studies to examine the separation performance of a fluidized bed containing sets of parallel inclined plates, as shown in Figure 7. The term Reflux Classifier has been used to describe these systems. Particles that fail to pass up through an inclined channel slide back down the incline, and are then given further opportunities to pass through. In hydrosizing, the inclined plates provide a significant throughput advantage by generating a higher hydraulic capacity. The inclined channels also provide a geometrical condition for determining whether a particle should or should not escape into the overflow. The system also offers a level of self-control. Feed fluctuations, in composition and rate, simply produce a change in the suspension concentration, which of course is free to vary due to the presence of the inclined plates. Thus, a fixed underflow rate can respond to load changes in particles that should report to the underflow. Naturally, some variation in hindered settling can occur if the disturbances result in excessive concentration changes.

Galvin et al. (2002) have also reported on the results of a pilot scale gravity separation trial using a vessel with cross-sectional area of 0.36 m². The solids feed rate was

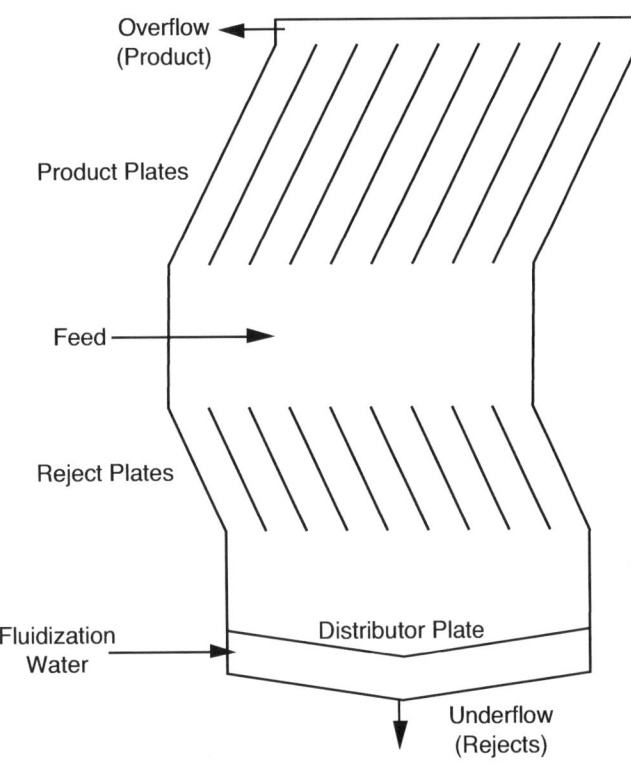

FIGURE 7 A schematic representation of the Reflux Classifier, consisting of a fluidized bed housing and two sets of parallel inclined plates

typically 15 t/h, which represents a loading more than three times higher than for a conventional system (Honaker and Mondal, 1999). Additional results from a typical separation of a coal and mineral matter feed are shown in Figure 8 using a set of partition curves. Each of the partition curves is for a different narrow size fraction of particles, and hence each individual curve is very sharp, with a relatively low E_p. The overall partition curve, which is a composite of the individual curves, is shown in Figure 9. In general our interest is in predicting the overall partition curve for a given feed. Given the E_p values of the individual curves are relatively small, it is reasonable to consider the separations of individual particle sizes as perfect, and then simply focus on the variation in the D_{50} with the particle size as this governs the overall E_p for the full size range.

Within the Reflux Classifier, the relevant separation condition is assumed to exist just above the sediment bed near the base of the vessel. Here the suspension density for the run reported in Figures 8 and 9 was 1,484 kg/m³. The interstitial fluid velocity in a steady state continuous separator is $U'/(1 - \phi)$, where U' is the net superficial velocity of liquid, equal to the fluidization liquid rate per unit area minus the underflow liquid rate per unit area. Here $U' \sim 2.6$ mm/s and the total volume fraction of solids just above the bed was ~0.4, assuming the relative density of the particles was about 2.2. Once again Equations 3, 10, and 20 were used to calculate the slip velocity. Figure 10 shows the slip velocity values for particle species ranging in size from 0.25 mm to 2.0 mm, and

34 | GRAVITY CONCENTRATION FUNDAMENTALS

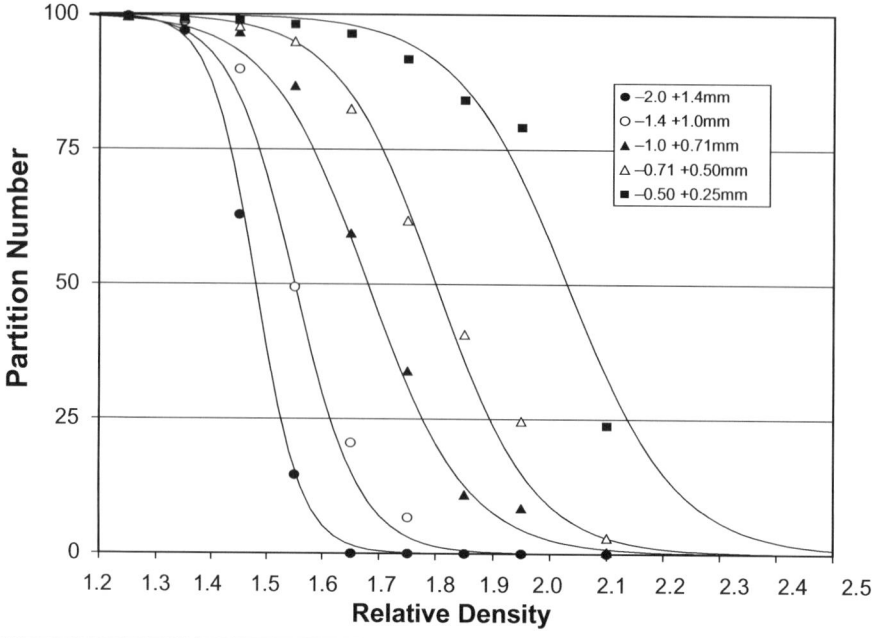

FIGURE 8 Partition curves for an energy coal processed using the pilot scale Reflux Classifier, obtained from the washability analysis of the feed, product, and reject streams

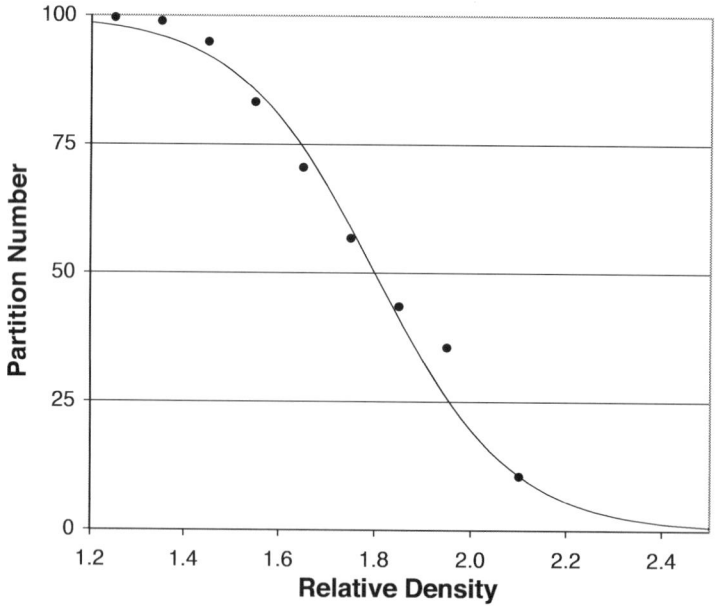

FIGURE 9 Overall partition curve based on the data in Figure 8 (D_{50} = 1.80, E_p = 0.16)

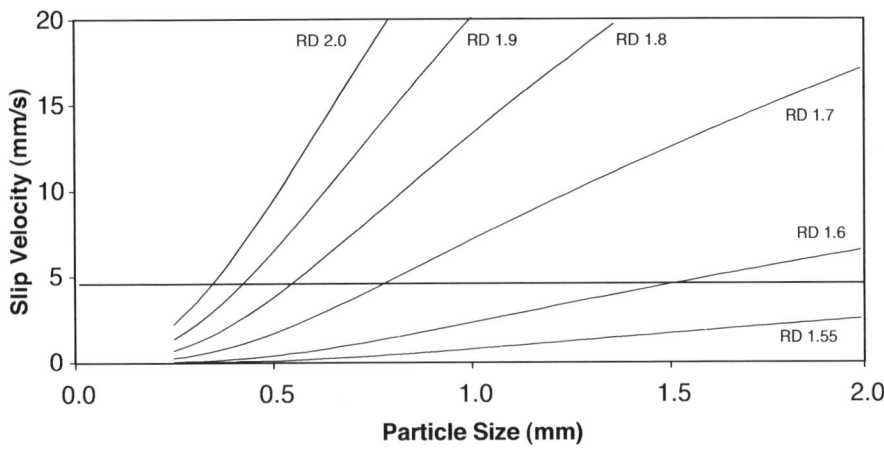

FIGURE 10 Slip velocity values calculated for a range of particle sizes and densities using a suspension density of 1,484 kg/m³. The horizontal line denotes the slip velocity of 4.3 mm/s at the principal separation point in the Reflux Classifier. The intersection with the curves provides the D_{50} values for the separation.

ranging in density from 1,550 to 2,000 kg/m³, with ρ_{susp} = 1,484 kg/m³. The horizontal line defines a special combination of particles having a common slip velocity, U_{slip} = U′/(1 − ϕ) = 4.3 mm/s. Figure 11 shows a plot of the particle density versus the particle size for a suspension density of 1,484 kg/m³ and slip velocity of 4.3 mm/s, based on Equations 3, 10, and 20. The square symbols denote the D_{50} values for the different particle size fractions shown in Figure 8. The triangles are based on a batch fluidization test of a feed produced by combining equal masses of the product and reject. The data were obtained using the second fluidization method described earlier, using a sample extracted from just above the packed bed at the base. The density of the suspension from the fluidized bed was 1,434 kg/m³. The agreement between the data sets is generally good, and consistent with that previously observed by Galvin et al. (1999b, 2002a).

CONCLUDING REMARKS

In this study we have reviewed a range of slip velocity models for describing the hindered settling that occurs as a result of the motion and the presence of neighboring particles. The emphasis of this paper has been on suspensions that contain a broad range of particle sizes and densities. In mineral processing, in particular, these suspensions are common, yet little work has been done to assess the adequacy of slip velocity models for such suspensions. A limited assessment of one slip velocity model was presented, covering a broad size and density range. Although the model performed reasonably well, there remains a need to extend the particle density range and to assess in more detail the effects of the actual size and density distributions that are present. The influence of dispersion on such systems also needs to be quantified.

This paper also outlined how fluidized beds can be used to quantify the density distribution of particles in a feed. These data, expressed in terms of the particle size distribution, are essential if predictions are to be made on the separation of a given feed. The paper also examined some recent developments in liquid fluidization, using inclined plates to increase segregation rates and hence influence the suspension concentration

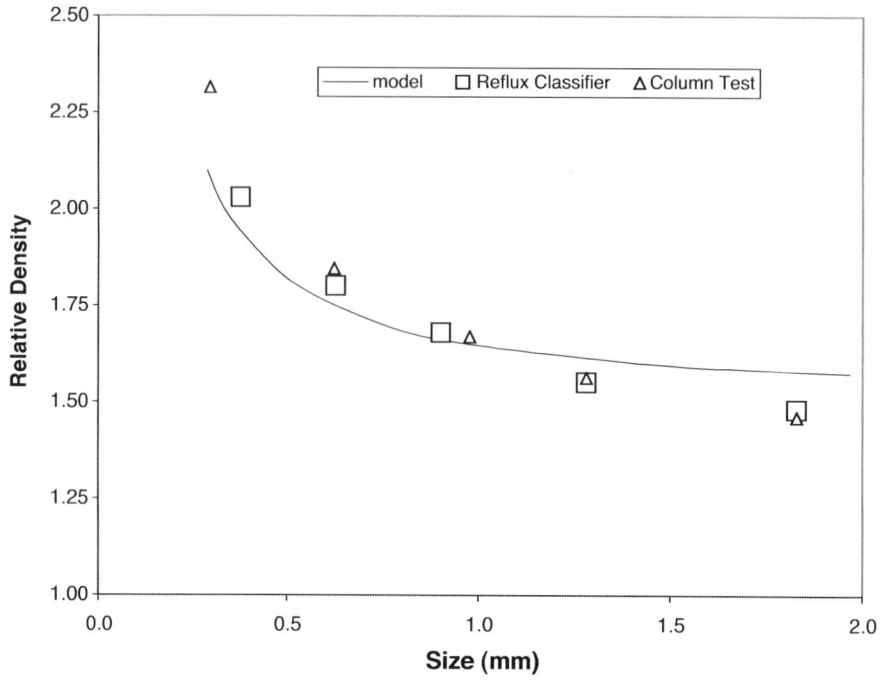

FIGURE 11 Relationship between the density and size of particles having a common slip velocity of 4.3 mm/s at a suspension density of 1,484 kg/m^3. The curve is based on Equations 3, 10 and 20, the squares denote the D_{50} versus size values from Figure 8, and the triangles denote results from a batch fluidized bed test based on a coal and mineral matter feed similar to that used in the Reflux Classifier.

and permit significantly higher solids throughput. This influence on the suspension concentration provides a form of self-control in size classification, and enhances the separation performance in gravity concentration.

ACKNOWLEDGMENTS

The author thanks the Australian Research Council, Australian Coal Association Research Program, and Ludowici Mineral Processing and Equipment for its support of the work conducted by Galvin and his co-workers.

REFERENCES

Al-Naafa, M.A., and Selim, M.S. 1992. Sedimentation of monodisperse and bidisperse hard-sphere colloidal suspensions. *A.I.Ch.E. Journal*, 38(10):1618–1630.

Asif, M. 1997. Modeling of multi-solid liquid fluidized beds. *Chemical Engineering Technology*. 20:485–490.

Asif, M. 1998. Segregation velocity model for fluidized suspension of binary mixture of particles, *Chem. Eng. Proc.* 37:279–286.

Asif, M., and Peterson, J.N. 1993. Particle dispersion in a binary solid-liquid fluidized bed. *A.I.Ch.E. Journal.* 39(9):1465–1471.

Barnea, E. and Mizrahi, J. 1973. A generalized approach to the fluid dynamics of particulate systems, Part 1. General correlation for fluidization and sedimentation in solid multiparticle systems. *The Chemical Engineering Journal.* 5:171.

Batchelor, G.K. 1972. Sedimentation in a dilute dispersion of spheres. *J. Fluid Mech.* 52. part 2. 245–268.

Batchelor, G.K. 1982. Sedimentation in a dilute polydisperse system of interaction spheres. Part 1. General theory. *J. Fluid Mech.* 119:379–408.

Batchelor, G.K., and Wen, C.S. 1982. Sedimentation in a dilute polydisperse system of interacting spheres. Part 2. Numerical Results. *J Fluid Mech.* 124:495–528.

Batchelor, G.K., and Green, J.T. 1972. The determination of the bulk stress in a suspension of spherical particles to order c^2. *J. Fluid Mech.* 56. part 3. 401.

Boycott, A.E. 1920. Sedimentation of blood corpuscles. *Nature.* 104:532.

Brauer, H., and Kriegel, E. 1966. Kornbewegung bei der sedimentation. *Chem. Ing. Tech.* 38(3):321–330.

Brauer, H., and Theile, H. 1973. Bewegung von partikelschwarmen. *Chem. Ing. Tech.* 45(13):909–912.

Brinkman, H.C. 1952. The viscosity of concentrated suspensions and solutions. *J. Chem. Phys.* 20(4):571.

Brodkey, R.S. 1967. *The phenomena of fluid motions.* 630. New York: Dover Publications.

Callen, A.M., Pratten, S.J., Belcher, B.D., Lambert, N., and Galvin, K.P. 2002a. Coal Washability analysis using water fluidization. submitted to *Coal Preparation.*

Callen, A.M., Pratten, S.J., Belcher, B.D., Lambert, N., and Galvin, K.P. 2002b. A new method for washability analysis of fine coal particles using water fluidization. *Australian Coal Preparation Conference.*

Callen, A.M., Pratten, S.J., Lambert, N., and Galvin, K.P. 2002c. Measurement of the density distribution of a system of particles using water fluidization. *4th World Congress on Particle Technology*, Sydney.

Clift, R., Seville, J.P.K., Moore, S.C., and Chavarie, C. 1987. Comments on buoyancy in fluidized beds. *Chem. Engng. Sci.* 42(1):194.

Davis, R.H., and Gecol, H. 1994. Hindered settling function with no empirical parameters for polydisperse suspensions. *A.I.Ch.E. Journal.* 40(3):570–575.

Di Felice, R. 1995. Hydrodynamics of liquid fluidisation. *Chem. Eng. Sci.,* 50(8):1213–1245.

Einstein, A., *Ann. Physik.* 1906. 19:289; *Ann. Physik.* 1911. 34:591.

Foscolo, P.U., Gibilaro, L.G., and Waldram, S.P. 1983. A unified model for particulate expansion of fluidised beds and flow in fixed porous media. *Chem. Engng. Sci.* 38:1251–1260.

Galvin, K.P., Pratten, S.J., and Nguyen-Tran-Lam, G. 1999a. A generalized empirical description for particle slip velocities in liquid fluidized beds, *Chem. Engng. Sci.* 54:1045–1052.

Galvin, K.P., Pratten, S.J., and Nicol, S.K. 1999b. Dense medium separation using a teetered bed separator. *Minerals Engineering.* 12(9):1059–1081.

Galvin, K.P., and Pratten, S.J. 1999c. Application of fluidisation to obtain washability data. *Minerals Engineering.* 9:1051–1058.

Galvin, K.P., Doroodchi, E., Callen, A.M., Lambert, N., and Pratten, S.J. 2002a. Pilot plant trial of the Reflux Classifier. *Minerals Engineering.* 15:19–25.

Galvin, K.P., Nguyentranlam, G. 2002b. Influence of parallel inclined plates in a liquid fluidized bed system. *Chem. Eng. Sci.* 57:1231–1234.

Garside. J., and Al-Dibouni, M.R. 1979. Particle mixing and classification in liquid fluidised beds. *Trans. IChemE.* 57:95–103.

Gibilaro, L.G., Felice, D.I., Waldram, S.P., and Foscolo, P.U. 1986. A predictive model for the equilibrium composition and inversion of binary-solid liquid fluidised beds. *Chem. Engng. Sci.* 41(2):379–387.

Honaker, R.Q., and Mondal, K. 1999. Dynamic modelling of fine particle separations in an hindered bed classifier. SME Annual Meeting, Denver, Colorado, preprint number 99-170.

Kennedy, S.C., and Bretton, J.F. 1966. Axial dispersion of spheres fluidized with liquids. *A.I.Ch.E. Journal.* 12:24–30.

Lockett, M.J., and Al-Habbooby, H.M. 1973. Differential settling by size of two particle species in a liquid. *Trans. Inst. Chem. Engrs.* 51:281–292.

Lockett, M.J., and Al-Habbooby, H.M. 1974. Relative particle velocities in two-species settling. *Powder Technology.* 10:67–71.

Masliyah, J.H. 1979. Hindered settling in a multi-species particle system. *Chem. Engng. Sci.* 34:1166–1168.

Moritomi, H., Iwase, T., and Chiba, T. 1982. A comprehensive interpretation of solid layer inversion in liquid fluidized beds. *Chem. Engng. Sci.* 37(12):1751–1757.

Nguyentranlam, G., and Galvin, K.P. 2001. Particle classification in the Reflux Classifier. *Minerals Engineering.* 14(9):1081–1091.

Ramirez, W.F., and Galvin, K.P. 2002. Mathematical model of segregation and dispersion in liquid fluidized beds. *to be submitted to A.I.Ch.E. Journal.*

Richardson, J.F., and Zaki, W.N. 1954. Sedimentation and fluidization: Part I. *Trans. Instn. Chem. Engrs.*, 32:35.

Zigrang, D.J., and Sylvester, N.D. 1981. An explicit equation for particle settling velocities in solid-liquid systems. *A.I.Ch.E. Journal.* 27(6):1043–1044.

Zuber, N. 1964. On the dispersed two-phase flow in the laminar flow regime. *Chem. Engng. Sci.* 19:897–917.

Modeling of Hindered-settling Column Separations

Bruce H. Kim*, Mark S. Klima*, and Heechan Cho[†]

A phenomenological model was developed to describe the operation of a hindered-settling column. The model is based on the convection-diffusion equation as applied to hindered-settling conditions. Simulations were carried out to evaluate column performance as a function of operating and design variables, including teeter water rate, bed height, solids feed rate, and column height. The results are presented in terms of fractional recovery (partition) curves. For selected tests, the simulation results are compared with experimental results obtained from the testing of a laboratory hindered-settling column.

INTRODUCTION

Hindered settling occurs when the settling rate of a particle in a liquid suspension is affected by the presence of nearby particles (Allen and Baudet, 1977). In a free-settling environment, the effect of particle size dominates the settling rate of particles, while during hindered settling, the effect of particle density on the settling rate is enhanced. The transition from free settling to hindered settling occurs when the concentration of solids in the suspension increases, reducing the distance between particles such that the drag force created by the settling particles affects the movement of nearby particles (Mirza and Richardson, 1979; Oliver, 1961). However, if the concentration of solids in the suspension is too high, entrapment and misplacement of particles will dominate, thereby increasing en-masse settling, which is independent of particle size and density (Davies, 1968).

Hindered-settling devices are used to separate particles based on differences in size and density. They have been used for separations in a wide range of industries, including aggregate (Rukavina, 1991), iron ore (Lutsky, 1999), and coal (Hyde et al., 1988). They

---

* Dept. of Energy and Geo-Environmental Engineering, University Park, Pa.

† School of Civil, Urban and Geosytem Engineering, Seoul National University, Seoul, Korea.

have also been used in environmental applications as a part of a soil washing system (EPA, 1992). Hindered-settling devices combine relatively low capital, installation, and operating costs with simple, fully automated operation (Hyde et al., 1988).

Various mathematical models have been developed for such size/density separation devices. For example, Kojovic and Whiten (1993) developed an empirical model for particles settling in a cone classifier, relating operating and geometric parameters and separation efficiency. Mackie et al. (1987) developed a hybrid physical-empirical separation-type model, which is based on settling theory, to describe the operation of a Stokes hydrosizer. Smith (1991) developed a mathematical model of an elutriator, using differential settling velocities of binary mixtures proposed by Lockett and Al-Habbooby (1973).

More recently, Honaker and Mondal (2000) developed a "dynamic population-balance model," to predict hindered-settling separations in a Floatex density separator. The model uses a hindered-settling equation developed by Brauer and coworkers to predict the hindered-settling velocities of particles (Brauer and Kriegel, 1966; Brauer, 1971; Brauer and Mewes, 1972; Brauer and Theile, 1973). Brauer's equation adjusts the free-settling velocities of particles in a suspension by applying two correction factors to account for the hindered-settling effect and, in turn, compute hindered-settling velocities. The first parameter accounts for an upward fluid flowing against the settling of a particle that is a function of the solids concentration. The second factor accounts for the particles settling in a dense suspension, resulting in variable flow profiles. The flow profile interactions produce turbulence, which was termed "cluster turbulence."

The settling velocities of particles are determined by integrating an equation for the particle acceleration, which is a function of the physical properties belonging to the particles and carrying fluid, over a given time interval. The computed velocities are adjusted for the hindered-settling condition using Brauer's hindered-settling equation, and these values are again integrated over the same time interval to identify the distance traveled by the particles. Based on specific characteristics associated with the physical and operating parameters, the domain of the Floatex density separator is divided into six distinct zones: overflow collection, upper intermediate, feed, lower intermediate, thickening, and underflow collection. The resulting movement of the particles between the zones is tracked by the population-balance method, which uses simple addition and subtraction of mass within each zone to calculate resulting mass accumulation. The results obtained with Honaker and Mondal's model show good agreement with their experimental data, which were obtained for coal cleaning separations.

Laboratory Hindered-settling Column

For the current study, the modeling approach was based on using the convection-diffusion equation under hindered-settling conditions. Figure 1 shows the schematic of the hindered-settling column used in this study. The design was based on that of a Linatex Hydrosizer™. The column was fabricated into three sections out of 8.25-cm (3.25-in) inside diameter Lexan tubing. The sections consist of the following: a 25.4-cm (10-in) long bottom section where the teeter water distributor and differential pressure sensor are located; a 25.4-cm (10-in) long top section with the overflow launder attached to it; and a 30.5-cm (12-in) long middle section. The length of the middle section can be changed to vary the overall column height. The column sits on a stainless steel dewatering cone, which is attached to an electro-pneumatic pinch valve. The valve is controlled by a PID controller, which also monitors the bed pressure. The teeter water and feed water rates are controlled with rotometers.

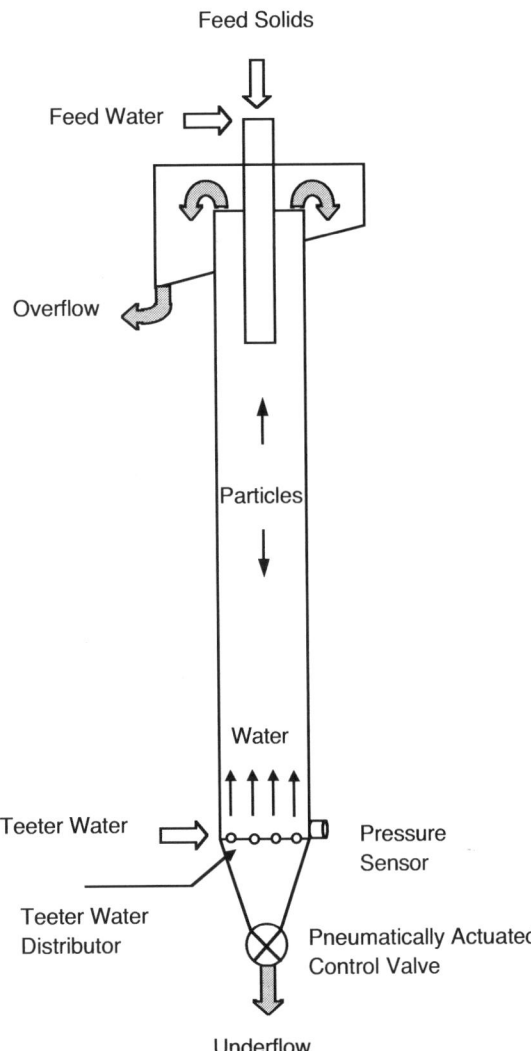

FIGURE 1 Schematic of the hindered-settling column

The solids are fed through the top opening of the column along with feed water. The function of the feed water is to prevent the solids from bridging in the feed tube, thereby ensuring a constant flow of feed solids into the column. Because of the upward flow of water (teeter water) in the column, the feed water is simply directed to the overflow stream, where light and finer particles are also collected. Because the feed water never enters the lower part of the column, it has no effect on the separation of solids.

The teeter water is injected through a distributor, which is located directly above the conical section (see Figure 1). The upward water velocity, which is a function of the teeter water rate and the cross-sectional area of the column, counters the downward settling velocities of the solids. Particles entering the column, which have a settling velocity that is less than that of the upward water velocity, are carried to the overflow stream.

The remaining solids settle in the column, building a high-density, fluidized bed. The height or thickness of the bed is measured using a differential pressure sensor (see Figure 1), which is monitored by the PID controller. The desired bed thickness is maintained by opening and closing a pneumatically-actuated pinch valve, which is controlled by the PID controller. The control system is essential in keeping the equilibrium balance inside the column between incoming feed solids and discharging products, as well as maintaining the proper fluidized bed density. Only particles having sufficient mass can pass through the bed and settle to the underflow stream. Thus, the fluidized particle bed creates an autogenous separating medium. Maintaining a constant density in the teeter bed is critical for achieving a successful separation of particles in the hindered-settling column (Hyde et al., 1988).

The key operating variables of the column are teeter water rate, solids feed rate, and pressure sensor set point. As noted previously, the teeter water rate is directly related to the upward velocity of the water current that counters the settling of the particles. Hence, it is an important variable in controlling the separation or "cut" point. The solids feed rate impacts the build up of solids in the bed. A low feed rate prevents the formation of an adequate bed, while a high feed rate can cause excessive compacting of the bed. The compaction prevents the bed from retaining a fluidized state, causing misplacement of fine particles to the underflow stream. The pressure sensor set point determines the desired equilibrium pressure (i.e., bed height) inside the column. For example, raising the set point increases the maximum solids concentration in the column, which changes the separation size and/or density in the column.

The purpose of this study was to develop a mathematical model for the hindered-settling column, which can be used as an analytical tool to evaluate separations under different operating and design conditions (Kim, 2002). Simulations were carried out to evaluate column performance as a function of teeter water rate, solids feed rate, set point, and column height. The results are presented in terms of fractional recovery (partition) curves. For selected tests, the simulation results are compared with experimental results.

MODEL DESCRIPTION

If particles are in motion in a column due to convection and diffusion, their movement within the device can be accounted for by using the convection-diffusion equation, which is given by

$$\frac{\partial \phi(x, \rho, z, t)}{\partial t} = D \frac{\partial^2 \phi(x, \rho, z, t)}{\partial z^2} - \frac{\partial (V(x, \rho, z, t)\phi(x, \rho, z, t))}{\partial z} \quad \text{(EQ 1)}$$

where $\phi(x, \rho, z, t)$ = volume fraction of particles of size x to $x + dx$ and density ρ to $\rho + d\rho$ in the element at z to $z + dz$ at time t; D = diffusion coefficient; $V(x, \rho, z, t)$ = velocities of particles of size x to $x + dx$ and density ρ to $\rho + d\rho$ in the element at z to $z + dz$ at time t, with respect to the wall of the device.

The first term in the right side of the Equation 1 represents the rate of accumulation in the element z to $z + dz$ due to diffusion, and the second term represents accumulation due to convection. According to the mass balance, the space vacated or occupied by departing or incoming particles, respectively, must be accounted for by the liquid medium, because this flow of particles occurs within the open liquid area. This implies that, if a given volume of particles moves in the z direction, then the same volume of liquid will have to move in the $-z$ direction. The rate of liquid in the element is given by

$$U_f(z, t)(1 - \phi(z, t))A$$

where $1 - \phi(z, t)$ = volume fraction occupied by the liquid in the element at z to z + dz at time t; $U_f(z, t)$ = velocity of the liquid in the element at z to z + dz at time t, with respect to the wall of the device; A = cross-sectional area of the element. The rate out of the liquid in the element is

$$U_f(z, t)(1 - \phi(z, t))A + \frac{\partial(U_f(z, t)(1 - \phi(z, t)))}{\partial z} A dz$$

The net rate of accumulation of liquid in the element is

$$\frac{\partial(U_f(z, t)(1 - \phi(z, t)))}{\partial z} A dz$$

Hence, by equating the net change of volume of solid to liquid, the rate of accumulation for the liquid in the same domain is given by

$$\frac{\partial(1 - \phi(z, t))}{\partial t} = -\frac{\partial(U_f(z, t)(1 - \phi(z, t)))}{\partial z} \quad \text{(EQ 2)}$$

Equation 1 requires an equation to predict the settling velocities of particles. In a free-settling environment, this is relatively simple to determine, for example, using Stokes or Newton's equations, depending on the particle Reynolds number. However, in a hindered-settling environment, in which particle settling velocities are a function of solids concentration, there are complex interactions between the particles and fluid (e.g., momentum transfer effect), which leads to the differential motion among particles of different sizes and densities relative to the fluid.

Hindered-settling Equation

Concha and Almendra (1979) developed an expression for the settling velocities of spherical particles settling in a multi-size, multi-density particle system. However, their equation did not account for a condition where particle density was less than that of suspension, which would allow particles to rise instead of settle. Using Concha and Almendra's hindered-settling equation as a starting point, Lee (1989) made several modifications and incorporated several empirical parameters based on statistical analysis to allow the bi-direction of particle settling. The final form of the hindered-settling equation is given as

$$U = \frac{20.52 \mu_f}{d \rho_p} f_1(\phi) \left\{ \left[1 + 0.0921 \left(\frac{d^3 |\rho_s - \rho_p| \rho_p g}{0.75 \mu_f^2} \right)^{1/2} f_2(\phi) \right]^{1/2} - 1 \right\}^2 \quad \text{(EQ 3)}$$

where

$$f_1(\phi) = \frac{(1 + 0.75 \phi^{1/3})(1 - \phi)(1 - 1.47 \phi + 2.67 \phi^2)^2}{(1 - 1.45 \phi)^{1.83}(1 + 2.25 \phi^{3.7})}$$

$$f_2(\phi) = \frac{(1 + 2.25 \phi^{3.7})(1 - 1.45 \phi)^{1.83}}{(1 + 0.75 \phi^{1/3})(1 - \phi)(1 - 1.47 \phi + 2.67 \phi^2)}$$

$f_1(\phi)$ and $f_2(\phi)$ are empirical functions that account for the effects of solids concentration (by volume), ϕ, on the settling velocities of particles of diameter d; μ_f = fluid viscosity; g = gravitational acceleration; ρ_s = particle density; ρ_p = pulp density and is calculated by

$$\rho_p = \rho_f(1-\phi) + \sum_{k=1}^{m} \rho_{s_k} C_k \qquad \text{(EQ 4)}$$

where ρ_f = fluid density; C_k = fractional volumetric concentration for a solid component k that has a unique combination of size x_k and density ρ_k; and m = total number of densities in the system. Lee (1989) showed that when using Equation 3, the differences between the observed and predicted values were generally within ±10%. The greater deviations tended to exist for higher solids concentrations, where more particle-particle interactions occur.

Solution Scheme

Because an analytical solution is not available for Equation 1, a finite-difference solution scheme was used (Austin et al., 1992). The following assumptions were also made: the wall effects were neglected; other than the hydrodynamic interactions, no other particle-particle interactions occurred; and the effects of particle collisions were negligible. The third assumption was discussed by Zimmels (1985). He stated that the attainment of terminal settling velocities following a collision might be considered nearly instantaneous if relaxation times are small. This implies that the system becomes locally non-uniform for only the brief period of collision, which is negligible compared with the time interval between collisions.

Figure 2 shows the breakdown of the hindered-settling column into elements as required for the finite-difference solution scheme. The height of the column, H, is divided by the total number of the elements, N, to give the length of the element, Δz. A particle system having a continuous size and density distribution is divided into M number of discrete components, each having a unique combination of an average particle diameter and an average density. These components are designated as k, for which k = 1, 2, 3, 4, ..., M.

To account for the particles that leave the system, two imaginary elements, UF (Element 0) and OF (Element N + 1), were added to the bottom and the top of the column, respectively. Once particles reach either of the imaginary elements, they remain there for the duration of Δt and disappear in the next time interval. The model simulation can be evaluated at any simulation time t by determining the fraction of particles entering Element UF versus Element OF. At Element F, feed solids are continuously added at each time interval.

The "set point" of the pressure sensor, which is the desired pressure level inside the column, is equivalent to the pressure gradient of height H of the suspension. Since pressure P is defined as the mass per unit area, the total mass inside the column of cross-sectional area A is obtained by multiplying P by A. It can be converted to the overall solids concentration inside the column by

$$PA = [(1-\phi')\rho_f + \phi'\rho_s]HA \qquad \text{(EQ 5)}$$

Solving for the set point concentration, ϕ', gives

$$\phi' = \frac{\frac{P}{H} - \rho_f}{\rho_s - \rho_f} \qquad \text{(EQ 6)}$$

This equation relates the set point value of the pressure sensor to the solids concentration.

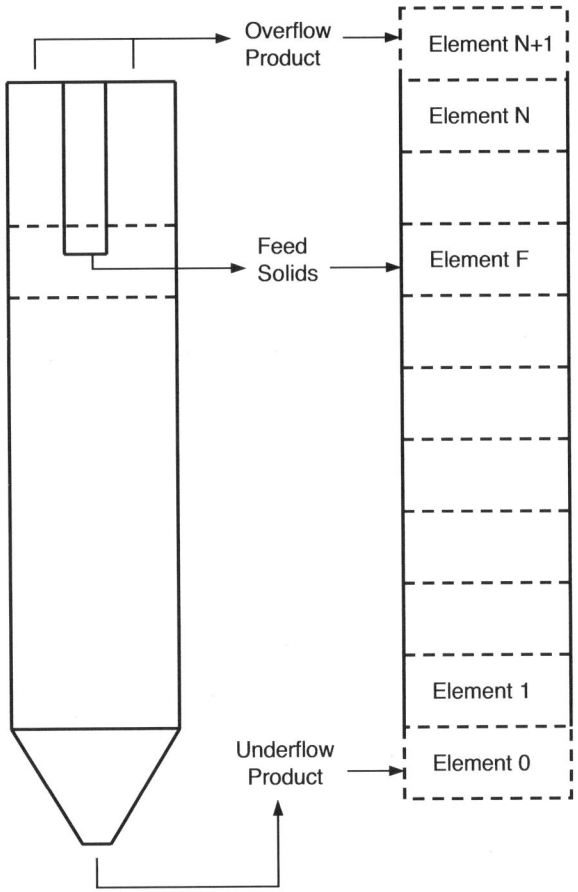

FIGURE 2 Representation of the hindered-settling column by elements as used in the finite-difference solution scheme

To incorporate the set point concentration into the hindered-settling model, it is important to include the operation of the control valve. In the early stages of the separation process, the control valve remains closed until a sufficient amount of dense and/or large particles build a bed that is thick enough to reach the set point concentration. From this point on, the actual overall solids concentration fluctuates around the set point concentration, and the control valve adjusts the opening size accordingly until an equilibrium condition is reached. This action is repeated through the entire time span of the column operation, while the overflow opening is left unregulated.

A FORTRAN program was written to solve what is referred to as the dynamic hindered-settling model (Kim, 2002). At every time interval, the program computes the overall solids concentration inside the column and compares it with the input value of the set point concentration. If the solids concentration is less than the set point concentration, the appropriate boundary condition is applied to Element 1, which simulates the closed underflow control valve. Similarly, if the solids concentration is greater than the set point concentration, the boundary condition is removed from Element 1, and the imaginary Element UF is added below Element 1.

Mass balances are determined for Element 1 and Element UF, which simulate the open underflow control valve. The whole process is repeated for every time interval to ensure that the overall solids concentration inside the column remains approximately equal to the set point concentration. When the desired simulation time is reached, the product and refuse are obtained. Since Δt is small, the closing and opening of the underflow valve are alternated very rapidly during the execution of the program, resembling a continuous process.

The rate of accumulation of mass into Element n from time t_1 to time t_2 is obtained by using the following finite difference solution equation

$$\frac{\Delta z(\phi_n^{t_2} - \phi_n^{t_1})}{\Delta t} = \frac{D}{\Delta z}(\phi_{n-1}^{t_1} - 2\phi_n^{t_1} + \phi_{n+1}^{t_1}) + V_{n+1/2}\left(\frac{\phi_n^{t_1} + \phi_{n+1}^{t_1}}{2}\right) - V_{n-1/2}\left(\frac{\phi_n^{t_1} + \phi_{n-1}^{t_1}}{2}\right) \quad (EQ\ 7)$$

where $V_{n+1/2}$ = settling velocity of particles at the interface between Element n and n + 1 relative to the wall of container; and $V_{n-1/2}$ = settling velocity of particle at the interface between Element n and n − 1 relative to the wall of container. Thus, the new solids concentration at time t_2 will be computed as

$$\phi_n^{t_2} = \phi_n^{t_1} + \frac{D\Delta t}{\Delta z^2}(\phi_{n-1}^{t_1} - 2\phi_n^{t_1} + \phi_{n+1}^{t_1}) + \frac{\Delta t}{\Delta z}\left[V_{n+1/2}\left(\frac{\phi_n^{t_1} + \phi_{n+1}^{t_1}}{2}\right) - V_{n-1/2}\left(\frac{\phi_n^{t_1} + \phi_{n-1}^{t_1}}{2}\right)\right] \quad (EQ\ 8)$$

Equation 8 is the basis of the mass balance between elements over time Δt and is repeated for each solid component k and Element n. For special elements such as Element 1 (changing boundary condition), Element N (top element), imaginary Elements OF and UF, and shock-front elements (discontinuities of solids), Equation 8 has to be modified in various forms to account for the conditions and to ensure the mass balance (Kim, 2002).

RESULTS AND DISCUSSION

The dynamic hindered-settling model is a complete model that integrates all the critical operating and physical parameters of the hindered-settling column, including solids feed rate, teeter water rate, feed solids composition, set point, column height, and feed location. For example, it is possible for solids to enter at different points along the column by adjusting the depth of the feed tube, and the separation process can be affected by this location as will be shown later. Other factors considered in the model include fluid density, fluid viscosity, diffusion, and gravitational acceleration.

One advantage of using the dynamic hindered-settling model is that it does not require an estimate of retention time. Rather, the "run" time is used, which is simply the measured time of the column operation before product samples are collected. By simulating the operation at different run times, it can be determined when steady-state operation occurs.

A series of simulations was performed using operating and design variables based on the laboratory hindered-settling column (see Figure 1). The baseline conditions were the following: column height = 76.2 cm (30 in); set point concentration = 23.12%; teeter water rate = 44.2 mL/sec (0.70 gal/min); feed rate = 1,200 g/min (2.64 lb/min); solids loading height fraction = 0.667; and run time = 600 sec. Limestone (2.73 g/cm^3 [170.3 lb/ft^3]) was used as the feed solids, and the size distribution is given in Table 1.

TABLE 1 Size distribution of the limestone (Young, 1999)

Sieve Size, µm	Percent Less Than Size
1,200	100
841	82.2
707	74.0
595	66.3
500	57.0
420	51.4
354	44.6
297	39.4
250	35.1
210	30.3
177	27.4
149	24.0
125	21.4
105	18.5
88	16.9
74	15.2
63	13.9
53	12.5
44	11.6
37	10.7
25	9.6

Fluid density, fluid viscosity, diffusion factor, and gravitational acceleration were 1.0 g/cm^3 (62.4 lb/ft^3), 0.01 g/cm/sec (2.42 lb/ft/hr), 6 cm^2/sec (23.25 ft^2/hr), and 981 cm/sec^2 (32.2 ft/sec^2), respectively, and remained constant for all simulations. The experimental results were based on those given by Young (1999; Young and Klima, 2000).

The results are presented in the form of fractional recovery (partition) curves as applied to size separations. They show the fraction of feed material of a particular size that will report to the coarse (underflow) stream. These curves can be used to estimate the separation or "cut" size of the separation. The cut size is defined as the particle size corresponding to a fractional recovery value of 0.5, i.e., a particle that has a 50% chance of reporting to the coarse stream.

Effect of Teeter Water Rate

Figures 3 and 4 are plots of fractional recovery curves for different teeter water rates. Figure 3 compares the individual experimental results with the corresponding simulation results for each teeter water rate, while Figure 4 shows the variation of the fractional recovery curves with teeter water rate for the experimental and simulated results, separately. As can be seen in Figure 3, good agreement was obtained between the simulated and experimental results. As expected, the cut size increased as the teeter water rate was increased (see Figure 4).

48 | GRAVITY CONCENTRATION FUNDAMENTALS

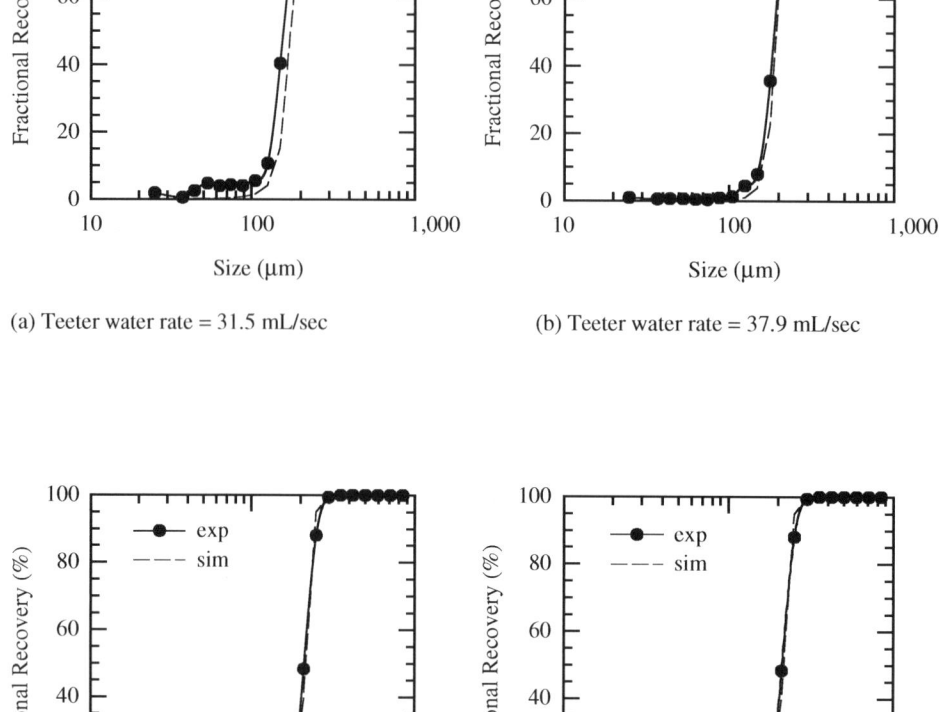

(a) Teeter water rate = 31.5 mL/sec

(b) Teeter water rate = 37.9 mL/sec

(c) Teeter water rate = 44.2 mL/sec

(d) Teeter water rate = 50.5 mL/sec

FIGURE 3 Fractional recovery curves comparing experimental and simulation results for various teeter water rates

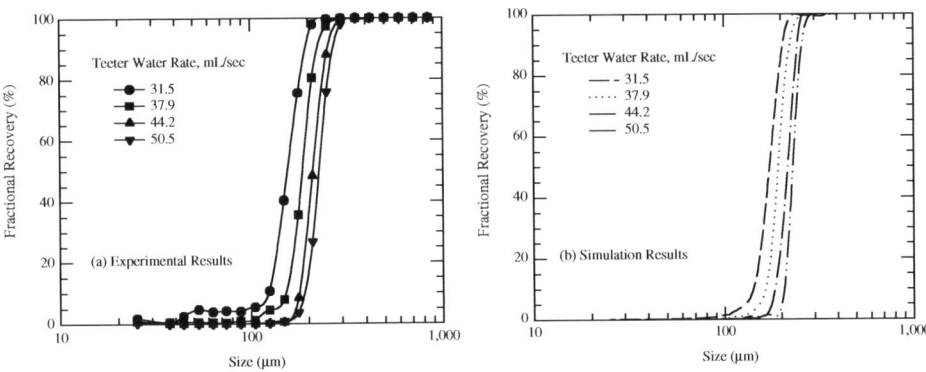

FIGURE 4 Variation of the fractional recovery curves with teeter water rate

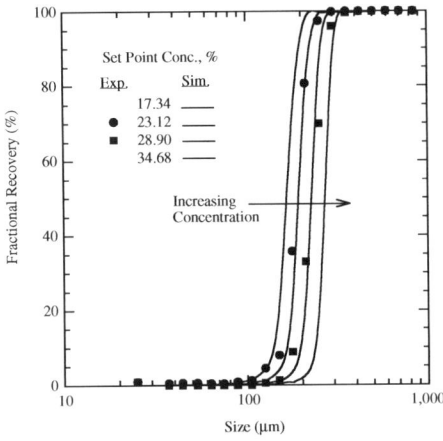

FIGURE 5 Variation of fractional recovery curves with set point concentration

Effect of Set Point Concentration

Figure 5 shows the variation of the fractional recovery curves with set point concentration (bed height) for both the experimental and simulated results. As the set point concentration was increased, the cut size also increased. This increase can be attributed to the denser bed produced at the higher set point.

Effect of Column Height

Figures 6 and 7 show the effects of column height on the fractional recovery curves. Again, the experimental and simulation results show good agreement. The cut size increased slightly as the column height decreased. The slight improvement in separation (somewhat steeper curve) for the taller column was likely due to the longer retention time, leading to less misplacement of particles. Overall, the performance was comparable at all heights even though the maximum column height was three times taller than the shortest one.

50 | GRAVITY CONCENTRATION FUNDAMENTALS

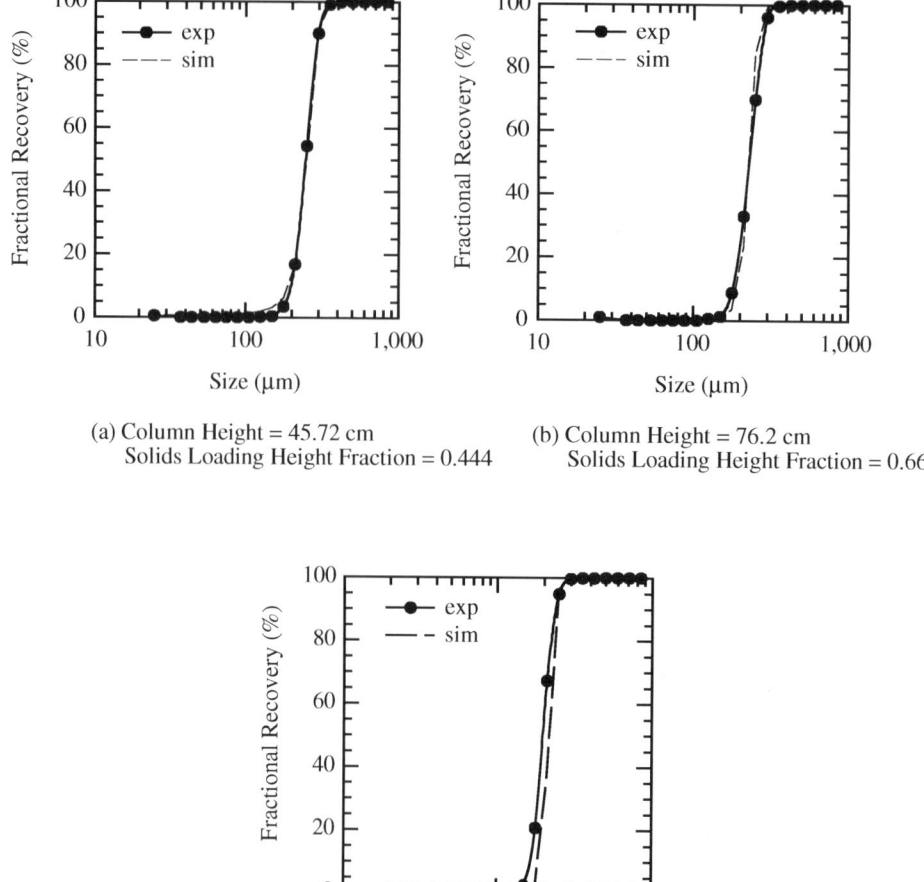

FIGURE 6 Fractional recovery curves comparing experimental and simulation results for various column heights

Effect of Solids Loading Height

The simulated fractional recovery curves for different solids loading heights are given in Figure 8. At either extreme, there is a greater probability that particles will be misplaced to the incorrect product. For example, at the top of the column, a large percentage of coarse particles are misplaced to the overflow stream, leading to a much coarser cut size and a poorer separation. On the other hand, feeding solids near the bottom of the column increases the probability that fine particles will be carried to the underflow stream. As can be seen, the cut size increases as the solid loading height fraction increases. In general, the solids loading height fraction of 0.5 produced the sharpest separation.

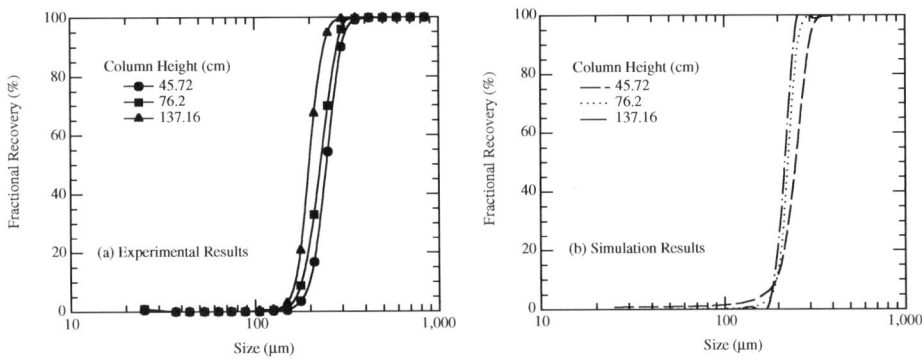

FIGURE 7 Variation of the fractional recovery curves with column height

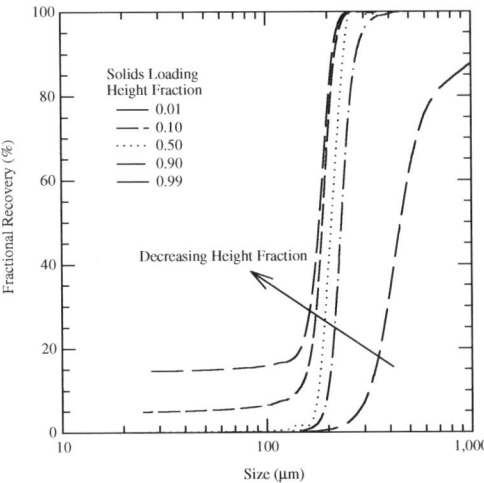

FIGURE 8 Variation of the simulated fractional recovery curves with solids loading height fraction

Effect of Solids Feed Rate

Figure 9 shows the simulated results for various solids feed rates, some of which were much higher than that tested in the laboratory column (i.e., 1,200 g/min). As can be seen, the separation is relatively unaffected by the change in solids feed rate until 2,400 g/min (5.29 lb/min). Above this value, the fractional recovery curves shift to coarser sizes (cut size increases), while flattening out, resulting in a lower separation efficiency. From an industry standpoint, a higher solids loading rate is beneficial, because it provides higher product output for a given time, and as was shown here, without sacrificing column performance. Moreover, the results also demonstrate that the column is able to handle large fluctuations in the solids feed rate, while maintaining a similar separation.

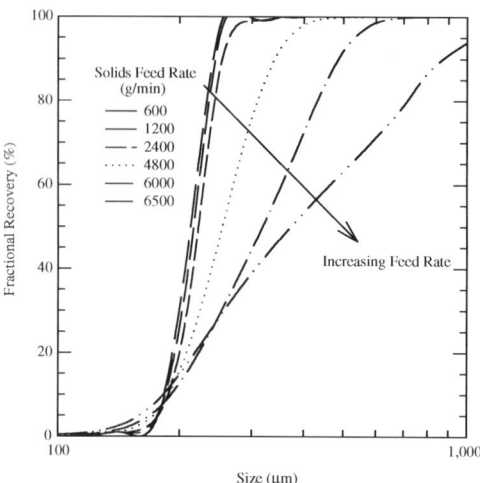

FIGURE 9 Variation of the simulated fractional recovery curves with solids feed rate

SUMMARY AND CONCLUSIONS

A mathematical model for the hindered-settling column was presented. The model was based on the convection-diffusion equation and incorporated an equation for calculating particle settling velocities under hindered-settling conditions. A finite-difference solution scheme was used to solve the convection-diffusion equation. The model was able to account for operating and design variables of the column including solids feed rate, teeter water rate, bed height, feed location, and column height.

The dynamic hindered-settling model provides an accurate representation of column performance. The simulated fractional recovery curves are in good agreement with the experimental results, leading to the following conclusions: an increase in teeter water rate or set point concentration produces a coarser product; shorter columns produce slightly coarser products, with only a slight decrease in column performance; column performance is relatively independent of solids feed rate over normal operating ranges; feeding solids near the top or bottom of the column should be avoided to prevent misplacement of particles to the improper product; while feeding at other points in the column has only a minor impact on column performance.

REFERENCES

Allen, T., and Baudet, M.G., 1977, "The Limits of Gravitational Sedimentation," *Powder Technology*, 18, 131–138.

Austin, L.G., Lee, C.H., Concha, F., and Luckie, P.T., 1992, "Hindered Settling and Classification Partition Curves," *Mineral and Metallurgical Processing*, 9, November, 161–168.

Brauer, H., and Kriegel, E., 1966, "Kornbewegung bei der Sedimentation (in German)," *Chem. Ing. Tech.*, 38, 321–330.

Brauer, H., 1971, Grundlagen der Einphasen- und Mehrphasen-stromungen (in German), Verlag Sauerlander, Aarau und Frankfurt/Main.

Brauer, H., and Mewes, D., 1972, "Stromungswiderstand Sowie Stationarer und Instatio-narer Stoff- und Warmeubergang an Kugeln (in German)," *Chem. Ing. Tech.*, 44, 865–868.

Brauer, H., and Theile, H., 1973, "Bewegung von Partikel-Schwarmen (in German)," *Chem. Ing. Tech.*, 45, 909–912.

Concha, F., and Almendra, E.R., 1979, "Settling Velocity of Particulate Systems, 2. Settling Velocity of Suspension of Spherical Particles," *International Journal of Mineral Processing*, 6, 31–41.

Davies, R., 1968, "The Experimental Study of the Differential Settling of Particles in Suspension at High Concentrations," *Powder Technology*, 2, 43–51.

EPA, 1992, "Soil/Sediment Washing System," *EPA/540/MR-92/075*, 1–2.

Honaker, R.Q., and Mondal K., 2000, "Dynamic Modeling of Fine Coal Separations in a Hindered-Bed Classifier," *Coal Preparation*, 21, 211–232.

Hyde, D.A., Williams, K.P., Morris, A.N., and Yexley, P.M., 1988, "The Beneficiation of Fine Coal Using the Hydrosizer," *Mine and Quarry*, 17(3), 50–54.

Kim, B.H., 2002, Modeling of Hindered-Settling Separations, Ph.D. Thesis, The Pennsylvania State University (in progress).

Kojovic, T., and Whiten, W.J., 1993, "Modeling and Simulation of Cone Classifiers," *XVIII International Mineral Processing Congress*, Sydney, 251–256.

Lee, C.H., Modeling of Batch Hindered Settling, Ph.D. Thesis, The Pennsylvania State University (1989).

Lockett, M.J., and Al-Habbooby, H.M., 1973, "Differential Settling by Size of Two Particle Species in a Liquid," *Trans. Instn. Chem. Engrs.*, 51, 281–292.

Lutsky, M., 1999, "The Use of Hydraulic Density Separators in the Iron Ore Industry," *SME Annual Meeting*, Abstract.

Mackie, R.I, Tucker, P., and Wells, A., 1987, "Mathematical Model of the Stokes Hydrosizer," *Transactions, Institution of Minerals and Metallurgy, Section C: Mineral Processing and Extractive Metallurgy*, 96, September, 130–136.

Mirza, S., and Richardson, J.F., 1979, "Sedimentation of Suspension of Particles of Two or More Sizes," *Chemical Engineering Science*, 34, 447–454.

Oliver, D.R., 1961, "The Sedimentation of Suspension of Closely-Sized Spherical Particles," *Chemical Engineering Science*, 15, 230–242.

Rukavina, M., 1991, "SIMI Develops a Recipe System," *Rock Products*, May, 36–41.

Smith, T.N., 1991, "Elutriation of Solids from a Binary Mixture," *Chemical Engineering Research and Design*," 69(5), 398–492.

Young, M.L., 1999, Evaluation of a Hindered-Settling Column for Size/Density Separations, M.S. Thesis, The Pennsylvania State University.

Young, M.L., and Klima, M.S., 2000, "Evaluation of a Hindered-Settling Column for Size/Density Separations," *Mineral and Metallurgical Processing*, 17, August, 194–197.

Zimmels, Y., 1985, "Theory of Density Separation of Particulate Systems," *Powder Technology*, 43, 127–139.

Dense Medium Rheology and Its Effect on Dense Medium Separation

Janusz S. Laskowski[*]

The yield stress in mineral suspensions strongly depends on particle size and solids content, and has a very strong stabilizing effect on fine mineral suspensions. Fine magnetite suspensions are characterized by high yield stress values and are therefore quite stable even at low medium densities. At higher medium densities the yield stress increases further making separation difficult. Because coarse magnetite suspensions are characterized by low yield stress values, they perform better at higher medium densities. It is beneficial to process very fine coal at higher inlet pressures using fine magnetite. Because of high yield stress associated with using fine magnetite, the magnetites with a bimodal particle size distribution give better separation results.

INTRODUCTION

The efficiency of gravity separation methods depends strongly on the size of treated particles and falls off rapidly for sizes finer than half of a millimetre. On the other hand, coal flotation also depends on particle size, and its efficiency significantly drops for coarser fractions (−0.5 +0.150 mm). Because the −0.5 +0.15 mm fraction generally contains the best quality coal, various flowsheets have been developed to treat these difficult size fractions.

In Canadian practice, the fine coal cleaning flowsheets include water-only cyclones and sieve bends prior to flotation. The popularity of a water-only cyclone is a result of its simple design and low cost/high capacity ratio. Single-stage installations operate with high probable errors which, however, can be significantly improved when cyclones are used in two-stage circuits (O'Brien and Sharpeta 1978).

A combination of classifying cyclones and modern spirals with flotation is typical for Australian plants. Since flotation results are affected by the petrographic composition of

---

[*] Dept. of Mining and Mineral Process Engineering, University of British Columbia, Vancouver, B.C., Canada.

coal and the size of particles (Klassen 1963; Firth, Swanson and Nicol 1978), it is also possible to improve flotation results by incorporating classification between the two stages of flotation to remove fine gangue particles from the rougher tailings prior to floating coarser coal in the second stage. Under such conditions the second stage flotation can easily separate coarser coal from coarser gangue particles, provided that it has its own conditioning tank into which some additional reagents are supplied. This was shown to remarkably improve the flotation of coarse coal particles (Firth, Swanson and Nicol 1979).

Among gravity separation methods, separation in dense media is the most efficient. Two types of dense medium separators are utilized to clean coal: A dense medium bath is used for coarse coal fractions while a dense medium cyclone is used to clean fine coal. The separation in a dense medium depends on density differences between light and heavy particles, their size and the properties of the medium. The separation efficiency strongly depends on the size of treated coal and the rheological properties of the magnetite dense medium. It is the relationship between the size of the treated coal particles and the rheology of the magnetite dense medium that is particularly important.

DENSE MEDIUM CYCLONES IN FINE COAL CLEANING

Southern Hemisphere coals are much more difficult to wash than coals from Europe and North America. The more difficult the coal is to wash, the more efficient cleaning methods are required, which makes selection of dense medium separation for difficult-to-wash coals obvious.

A two-stage dense medium cyclone (DMC) circuit was installed in the colliery situated near Witbank (South Africa) in 1971. The plant treated +0.5 mm feed and produced a 7% ash metallurgical concentrate and 15% ash middlings. In 1975, the float-sink tests carried out with the –0.5 mm material from this colliery showed that, at a cut point of about 1.48, a theoretical yield of 58% for a product ash of 7% should be obtained (Van der Walt, Falcon and Fourie 1981). At this cut point, the yield of the ±0.1 g/cm^3 near density material of about 40% indicated, however, that only dense medium cyclones could be realistically considered. Van der Walt et al (1981) reported that remarkably sharp separations were achieved almost immediately in the 5 t/h dense medium pilot plant while processing the –0.5 +0.075 mm coal and that there was no difficulty in obtaining a reasonable yield of 7% ash coal. They also observed that the main problem was magnetite. For good separation, the magnetite was to be ground finer than usual (to about 50% minus 10 μm) making magnetite recovery difficult.

The first commercial two-stage dense medium plant to process –0.5 +0.1 mm coal was built at Greenside Colliery. The underflow from the 1st stage DMC after dilution with water is pumped to the second stage DMC washing for separation into middlings and discard. As pointed out by the authors, the plant started up smoothly and produced the desired quality of products virtually from the outset. More recent data indicate that the plant, which is still in operation, uses the magnetite ground to 93% below 45 μm, and operates at an organic efficiency in excess of 90%. This compares favorably with the organic efficiency of about 80% reported for the South African spiral plants (Harris and Franzidis 1995).

A very interesting development in this area took place in Australia. The original processing plant at the Curragh Mine utilized two-stage dense medium cyclones to process a –32 +0.5 mm fraction of the feed. The –0.5 +0.075 mm fines were processed in conventional flotation cells to produce a coking coal concentrate, while the –0.075 mm material was discarded along with tailings. Since the flotation plant was needed to recover saleable

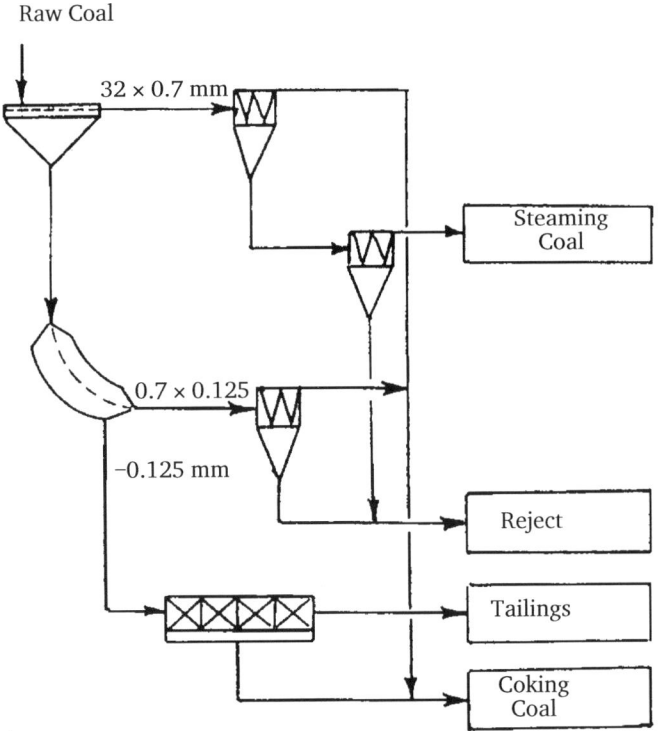

FIGURE 1 Curragh Mine coal preparation plant flowsheet

coal from the sub 0.075 mm fraction, a second DMC circuit was added to treat the −0.5 +0.075 portion of the feed. The flowsheet of the Curragh Mine plant which was in use for a couple of years is shown in Figure 1 (Kempnich, van Barneveld and Lusan 1993).

It is to be pointed out that as early as 1963, various pilot plant trials on fine coal cleaning in dense medium cyclones indicated that higher inlet pressures and finer magnetite were better for separation (Geer and Sokaski 1963). Stoessner et al (Stoessner, Chedy and Zawadzki 1988) reported that in the tests on cleaning the −0.6 +0.15 mm coal in a dense medium cyclone at low separation densities (1.3 g/cm^3) the sharpest Tromp curves were obtained when using the finest magnetite.

The Curragh Mine results confirmed that magnetite particle size distribution is critical in maintaining medium stability and dense medium cyclone performance. An important feature of the plant was the ability to control the magnetite size distribution. This was achieved by bleeding the reject drain medium out of the fine coal circuit and replacing it with make-up medium taken from the coarse coal dense medium cyclone overflow return. These measures, along with improved Climaxx counter-rotational magnetic separators, provided a continuous replacement of coarse magnetite with fine magnetite and allowed the fine coal circuit magnetite size distribution to be maintained. Another important feature was a much higher inlet pressure (the relationship between the inlet pressure and fineness of the magnetite is another very important relationship which will be discussed separately).

Dense medium cyclones are low-pressure devices and traditionally work at a standard inlet pressure of $9 \cdot D \cdot \gamma$ (where D is the diameter of the cyclone, and γ is the medium

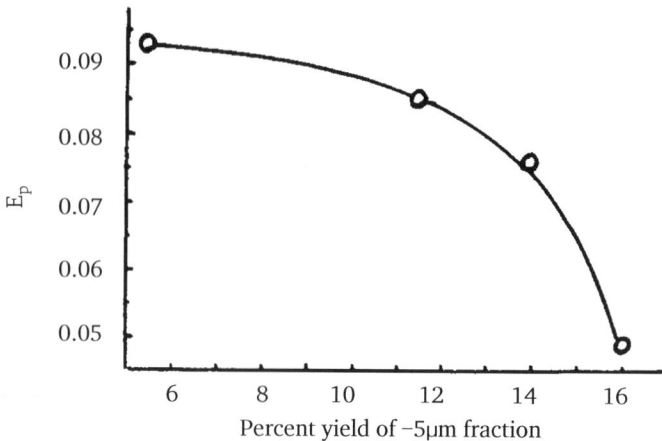

FIGURE 2 Effect of the content of –5 µm fraction of magnetite on separation efficiency of fine coal at low separation densities

density) which for the Curragh mine DMC was about 45 kPa. An increase in the inlet pressure to 65 kPa improved the Ep in the +0.25 mm fraction but the performance on the –0.25 mm material remained poor. However, a further increase in the inlet pressure to 120 kPa improved efficiency and from this point the overall target product ash below 7% was consistently obtained. Low Ep's of 0.04–0.05 were achievable on the +0.25 mm fraction and Ep's in the range of 0.1–0.11 were achievable on the –0.25 mm fraction. These separation efficiencies are obviously much better than the efficiencies of other separating devices.

MAGNETITE FOR DENSE MEDIUM PREPARATION

Magnetite Particle Size Distribution

As pointed out, magnetite size distribution plays a very important role in fine coal cleaning in DMC's. Figure 2 shows the effect of the –5 µm magnetite on the separation efficiency of –0.6 +0.015 mm coal at low separation densities (1.3 g/cm^3)(Stoessner, Chedgy and Zawadzki 1988). Increasing the percentage yield of –5 µm magnetite from 6% to 16% caused the Ep to decrease from 0.09 to 0.05. In spite of that, various industrial standards do not specify particle size distribution of magnetite precisely enough.

British coal mining industry standards specify that the medium should not contain more than 5% by mass of particles larger than 45 µm and 30% finer than 10 µm (Mikhail and Osborne 1990). Australian magnetite which comes from Biggeden and Tallawang mines is offered as a superfine magnetite (92–95% passing 53 µm) and an ultrafine magnetite (95–99% passing 53 µm) (Robertson and Williams 1991). US standards for A magnetite require 65% particles finer than 45 µm, and for B magnetite 90% finer than 45 µm. EXPORTech Co. is more precise; it offers commercial grade magnetite that is 95% finer than 45 µm and contains 15% of particles finer than 5 µm.

As these examples indicate, in most cases only top particle sizes are specified. This may indicate "regular" RRB (Rosin-Rammler-Bennett) distributions. But is it enough to specify the top particle size only?

TABLE 1 Particle size distribution of the tested magnetite samples

Sample	$d_{63.2}$ (μm)	m
Magnetite #1	30.5	3.2
Magnetite #2	18.0	1.6
Magnetite #4	4.3	1.9
Magnetite #5	2.7	2.5

The US Department of Energy developed the micromag, micronized magnetite that may contain either 70%, 80% or 90% particles below 5 μm (Klima, Killmeyer and Hucko 1990). Practically, these products are all 100% below 10 μm. Obviously, such a fine magnetite will be much more expensive than the magnetite used thus far commercially in the dense medium cleaning of coal. The question then arises, What is the optimum magnetite particle size distribution?

Rheology of Magnetite Dense Medium

The particles treated in a dense medium either sink or float. The separation efficiency is determined by the rate of movement of the treated particles in the medium. This movement, and especially movement of near-density particles, depends on medium viscosity and it is clear that various questions regarding separation efficiency and the effect of the medium can be answered only via rheological measurements.

Measurements of rheological properties of mineral suspensions are not easy. Usually such suspensions are unstable and settle with time. It has been known that magnetite suspensions in water exhibit properties of non-Newtonian fluids (Berghofer 1959). More recent measurements were carried out with a specially developed viscometer (Klein, Partridge and Laskowski 1990; Klein, Laskowski and Partridge 1995) using magnetites with various particle size distributions (Table 1).

The size and distribution moduli in Table 1 are those of the Rosin-Rammler-Bennett particle size distribution function:

$$F(d) = 100\left(1 - \exp\left(-\frac{d}{d_{63.2}}\right)^m\right) \quad \text{(EQ 1)}$$

where F(d) is the cumulative percent passing on size d, $d_{63.2}$ is the size modulus (it is that aperture through which 63.2% of the sample would pass), and m is the distribution modulus (the slope of the curve on RRB graph paper).

Magnetite #1 was a commercial grade magnetite provided by Craigmond Mines; this magnetite is commonly utilized in dense medium separation in Western Canada. Mag #2 was prepared by grinding Mag #1 in a ball mill. Mag #4 and Mag #5 were the micronized-magnetites kindly provided by the U.S. Department of Energy, Pittsburgh (Mag #4 is 70% below 5 μm, Mag #5 is 90% below 5 μm). In this project Magnetite #3 was also tested; this sample was prepared by classification of Magnetite #1 and rejection of fine fractions.

Figure 3 shows the rheological curves for the tested magnetite suspensions for volumetric magnetite contents from 5% to 30% (the magnetite medium density range of 1.3 to 1.7 corresponds to the volumetric magnetite contents from 5 to 20% vol).

As Figure 3 demonstrates, the flow curves exhibit shear-thinning properties with yield stress. The differences between the yield stress values are obviously very different

60 | GRAVITY CONCENTRATION FUNDAMENTALS

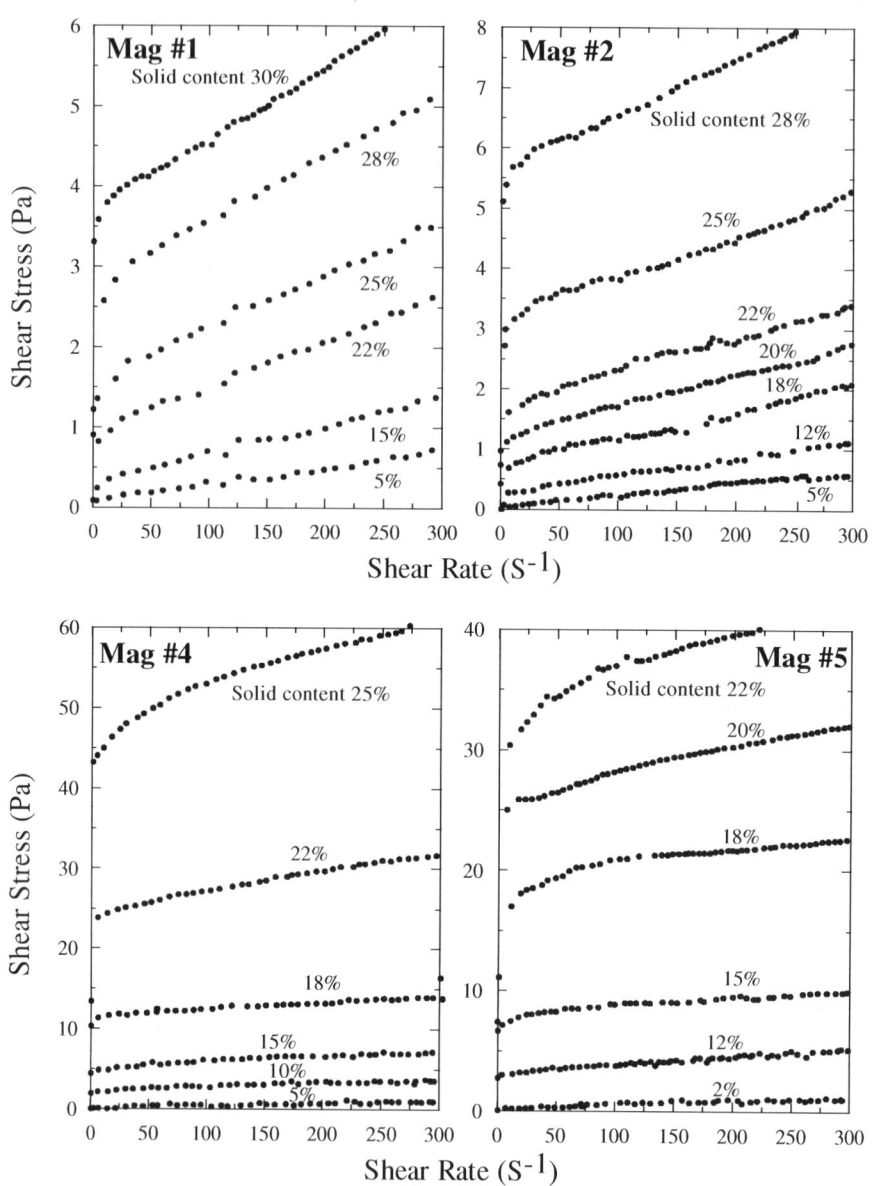

FIGURE 3 The flow curves of magnetite suspensions (solid content is % vol.)

for the tested magnetites (please note different scales in Figures 3a–3d). The rheological curves shown in Figure 3 were fitted to five models (Table 2) (He and Laskowski 1999).

Three models (the Bingham, the Herschel-Bulkley and the Casson) were found to describe the experimental flow curves quite well. The Casson model turned out to be very effective in describing the flow curves in a broad range of magnetite down to a micronized magnetite with $d_{63.2} = 2.7$ µm, and over the solid content range from 5% to 25% by volume (He and Laskowski 1999). Thus, the medium rheology over the above

TABLE 2 The tested rheological models

Name	Model
Newton	$\tau = \eta D$
Bingham	$\tau = \tau_{pl} + \eta_{pl} D$
Ostwald	$\tau = KD^n$
Herschel-Bulkley	$\tau = \tau_o + KD^n$
Casson	$\tau = [\tau_c^{1/2} + (\eta_c D)^{1/2}]^2$

Symbols used in Table 2: τ stands for yield stress, η for dynamic viscosity, D for shear rate, τ_{pl} for plastic (Bingham) yield stress, η_{pl} for plastic viscosity, τ_c for Casson yield stress and η_c for Casson viscosity.

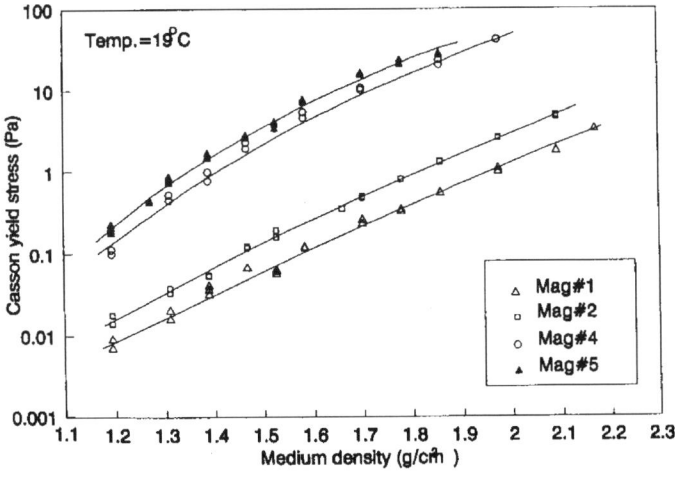

FIGURE 4 The effect of magnetite particle size and medium solid content on Casson yield stress

range can be described by two rheological parameters: Casson viscosity, η_c, and Casson yield stress, τ_c.

Figure 4 illustrates the effect of solid (magnetite) content for the four tested magnetite samples on the Casson yield stress.

As Figure 5 shows, the Casson viscosity is not very sensitive to the changes in magnetite particle size or solid content over broad medium density ranges. Only the Casson yield stress responds to these changes in a logical and well-defined manner (Figure 4). It is to be pointed out that although these two parameters are independent, they jointly determine the rheology of magnetite suspensions. For example, the apparent viscosity of such a system is given by (He and Laskowski 1996; He, Laskowski and Klein 2001):

$$\eta_a = \left(\sqrt{\frac{\tau_c}{D}} + \sqrt{\eta_c} \right)^2 \quad \text{(EQ 2)}$$

Since, as shown in Table 3, the Casson yield stress values are many times higher than those of the Casson viscosity, the effective viscosity of the medium is practically determined by the medium yield stress.

TABLE 3 Casson yield stress and viscosity for four grades of magnetite suspensions with a medium relative density of 1.45

Sample	τ_c (mPa)	η_c (mPa.s)
Mag#1	62	1.85
Mag#2	118	1.50
Mag#4	2,110	0.70
Mag#5	2,660	6.69

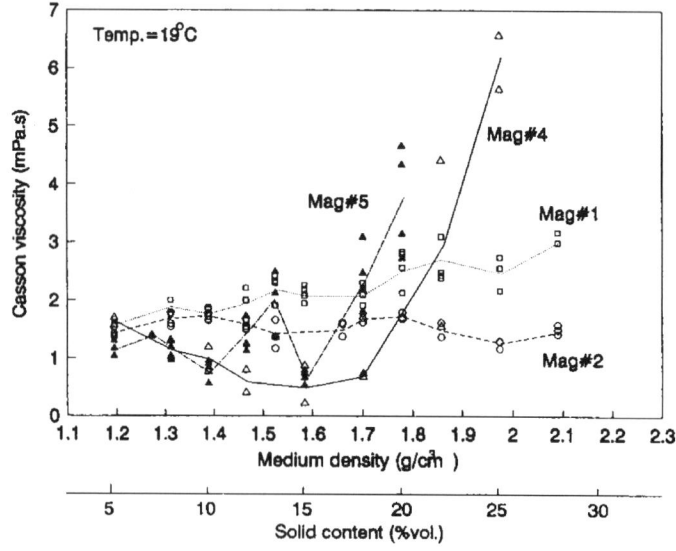

FIGURE 5 The effect of magnetite size and medium solid content on Casson viscosity

In a dense medium cyclone, the magnetite medium particles, like the coal particles, are also subjected to the centrifugal forces. This effect results in differences in densities between cyclone overflow and underflow which characterize the medium stability under dynamic conditions.

As Figure 6 reveals, these differences may be very large for coarse magnetite suspensions (Magnetites #1 and #3). For all magnetites the density differential decreases at higher solids content because of the stabilizing effect of increasing yield stress (Figure 4). These tests were carried out in a pilot scale DMC loop (He and Laskowski 1994) with a 150 mm (6") Krebs cyclone (model D6B-12-S287) at a 10 × D inlet pressure. To test the effect of magnetite medium properties on separation efficiency, color-coded density tracers obtained from Partition Enterprises Ltd., Australia, were used as cyclone feed. Three narrow size fractions were utilized in the tests: 4 × 2 mm, 1.0 × 0.71 mm and 0.5 × 0.355 mm.

Bimodal Magnetite Particle Size Distributions

As is known, suspensions with bimodal particle size distributions exhibit much lower viscosities than the suspensions with monomodal particle sizes (Farris 1968; Barnes, Hutton and Walters 1989). This effect is especially pronounced at a high solids content. Our experiments with a magnetite dense medium (density of 1.55 g/cm^3)

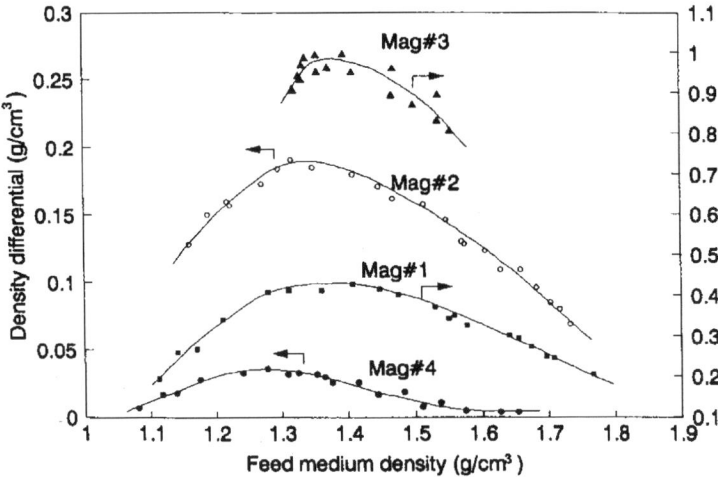

FIGURE 6 Effect of medium composition on density differential (at inlet pressure of 10 × D)

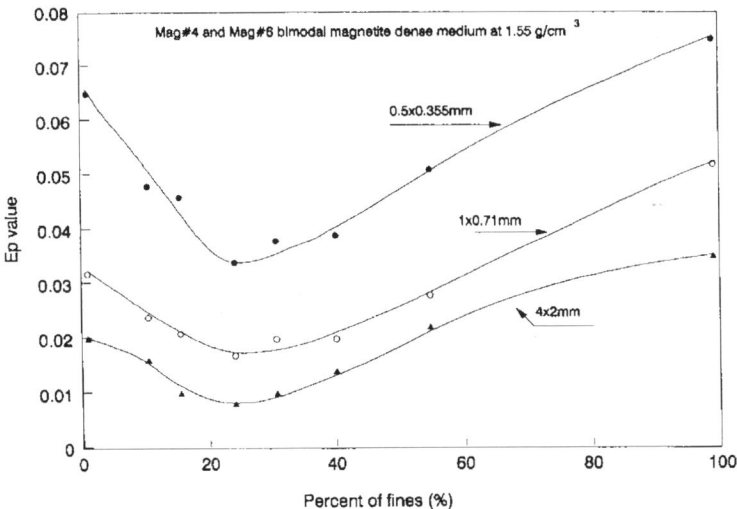

FIGURE 7 Separation efficiency as a function of the proportion of fines in the bimodal dense medium (at inlet pressure of 10 × D)

prepared by mixing Magnetite #4 (Table 1) with Magnetite #3 gave very promising results (Figure 7) (He and Laskowski 1995b).

The tests carried out with the medium which contained 30% of fine Magnetite #4 and 70% of coarse Magnetite #3 gave much lower Ep values.

Additional rheological tests were carried out in order to study further the effect of the solid content and magnetite particle size distribution on rheology of magnetite suspensions. In these tests Magnetite #3 was used in combination with the very fine Pea Ridge

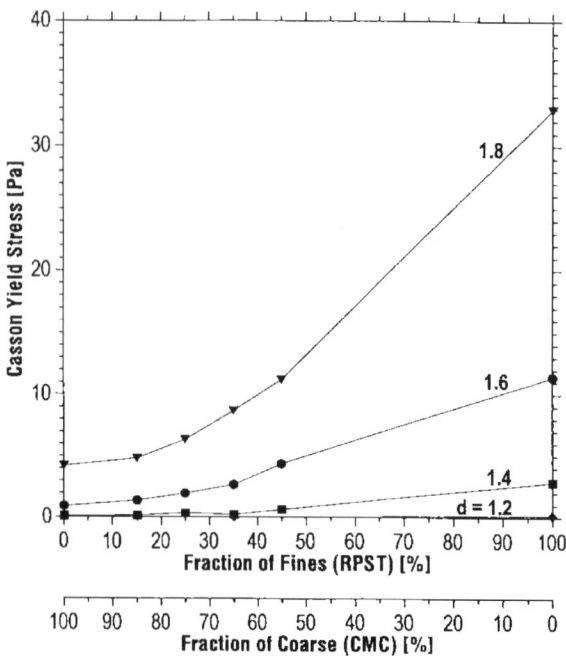

FIGURE 8 The Casson yield stress as a function of the proportion of fine (RPST) in bimodal medium with Mag #3 at different medium densities

RPS magnetite sample which was further ground in a stirred ball mill down to $d_{63.2} \approx$ 5 μm (designated RPST). The RPST magnetite sample was very similar to Magnetite #4 (Table 1). The Casson yield stresses determined for such bimodal magnetite samples (combination of RPTS and Magnetite #3 at various ratios) are shown in Figure 8. It confirms that the solid content (medium density) has a very pronounced effect on the measured yield stress. Up to about 35% of fine magnetite fraction, the yield stress stays low irrespective of the medium density (magnetite content) and increases only when the yield of fines exceeds this value. The measured apparent viscosity at 100 1/s is shown in Figure 9. In increases only at fine RPST magnetite content higher than 35%. This agrees remarkably with Equation 2 that gives the effect of yield stress on apparent viscosity.

The magnetite particle size distribution has a strong effect not only on the medium rheology but also on dynamic stability of the medium in DMC. Figures 10a and 10b show the effect of inlet pressure on the dynamic medium stability for commercial Magnetite #1 and the BMT35 bimodal sample (35% RPTS and 65% Magnetite #3). For the BMT35 bimodal sample, the increasing inlet pressure affects only very weakly the density differential. Of course, for the very fine RPST the medium is very stable irrespective of the inlet pressure (not shown here). As Figure 6 shows, at any medium density the density differential increases with increasing magnetite particle size. With increasing medium density, the density differentials for all the tested magnetite dense media became initially greater and fell off at high medium densities. Over the higher density range, the medium yield stress becomes significant, and this stabilizes the system. It is also to be pointed out that the density differentials reported in Figure 6 were measured at a 10 × D inlet pressure. As Figure 10a indicates, the density differentials become

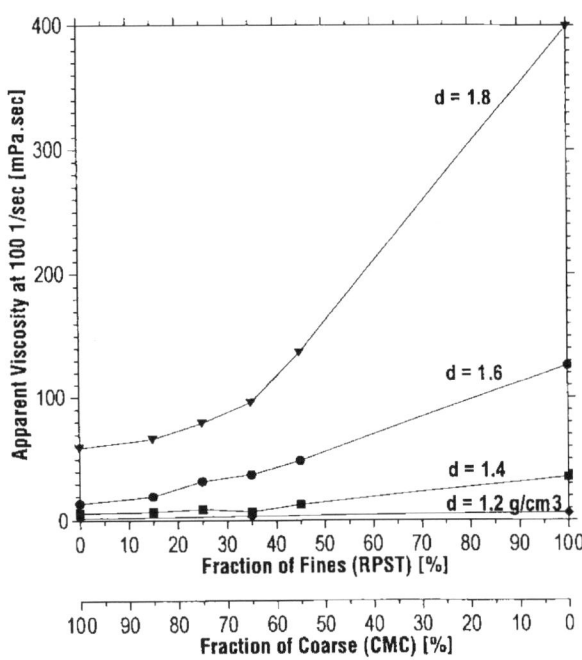

FIGURE 9 Apparent viscosity at shear rate of 100 1/s as a function of the proportion of fines (RPST) in the bimodal medium (mixture with Mag #3) at different medium densities

larger at increasing inlet pressures but this trend is somehow reduced for bimodal suspensions (Figure 10b).

These figures seem to explain contradicting reports on the effect of magnetite particle size on separation efficiency. While for finer magnetite the yield stress increases making separation in the dense medium more difficult, the dynamic stability of the medium is improved. The latter is especially true for higher inlet pressures which were claimed to improve separation efficiency. Again this may be true for fine magnetite but it does not hold true for a commercial grade magnetite. So, any increase in separation efficiency at higher inlet pressures may require finer magnetite which, however, will be characterized by higher yield stress values and higher effective viscosity.

At low inlet pressure, the dense media prepared from Magnetites #1, #2 and #4 are quite stable under dynamic conditions, and as Figure 11 shows the separation efficiency at low densities is good. For Mag #4, because of the high yield stress (high effective viscosity), the separation efficiency at densities higher than 1.5 sharply decreases. Viscosity of the Mag #3 medium could not be experimentally measured due to extremely low stability of this system. As Figure 6 shows, the stability of the medium prepared from Mag #3 is low, but because of increasing viscous properties at higher solid content, the stability increases at higher densities. The same trend is observed in Figure 11. The separation efficiency for Mag #3, which is low at low densities, improves at higher densities.

It can also be observed from the slope of the Ep versus medium density curve (Figure 11) that the separation efficiency is also a function of magnetite particle size. For finer magnetite the Ep values increase more rapidly at higher densities than for coarser magnetite media.

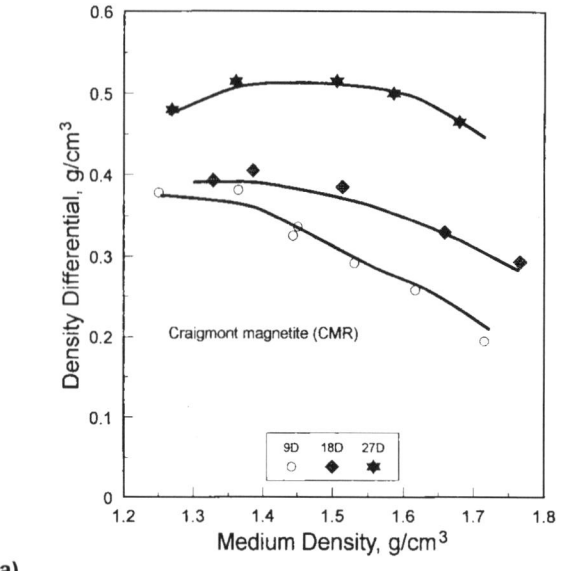

FIGURE 10 (a) Density differential as a function of feed medium density for regular Craigmont Magnetite (Mag #1) at various inlet pressures. (b) Density differential as a function of feed medium density for BMT35 Magnetite at various inlet pressures.

DENSE MEDIUM RHEOLOGY | 67

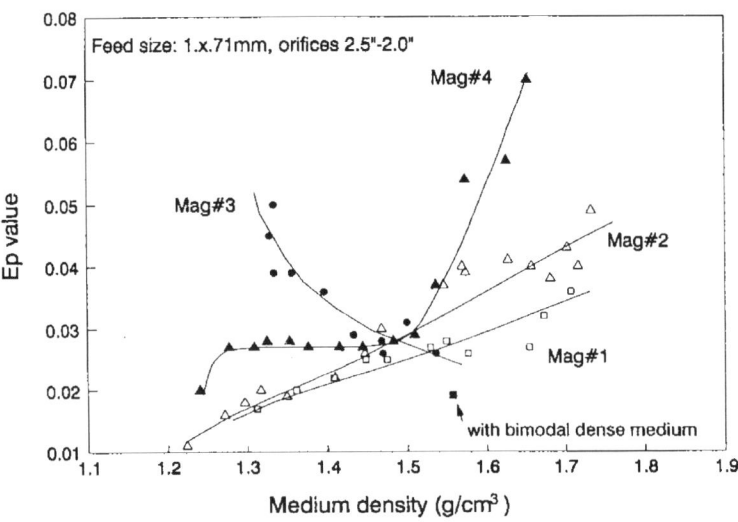

FIGURE 11 Effect of medium composition on DMC separation efficiency (at inlet pressure of 10 × D)

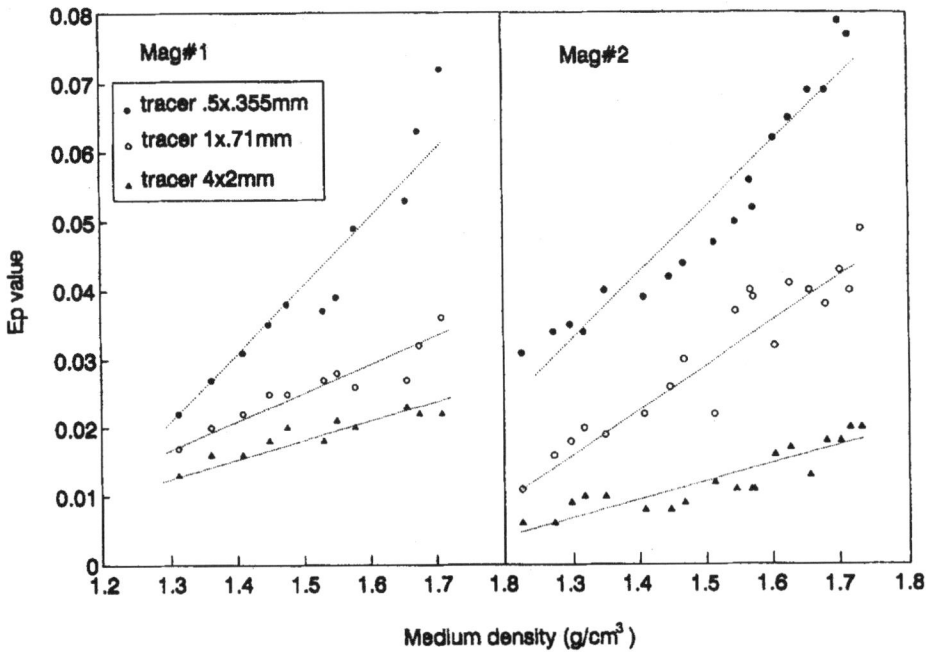

FIGURE 12 DMC separation efficiency as a function of medium density and magnetite particle size distribution (at inlet pressure of 10 × D)

The DMC performance also depends on the feed particle size. As seen from Figure 12, with finer feed (tracer) particles, the Ep values increase more rapidly especially over the high medium density range. The relationship between Ep value and medium density may be roughly approximated as linear (He and Laskowski 1995). The slope may be used as a parameter to evaluate the sensitivity of the separation to changes in medium density.

Collins et al (1983) reported that the separation efficiency could be increased by decreasing the density differential (that is by increasing magnetite suspension stability) from 0.85 to 0.4 g/cm^3. However further decrease in the density differential from 0.4 to 0.1 g/cm^3 was not beneficial for the separation efficiency. This indicates that the DMC separation requires the existence of a given density gradient inside the cyclone. It is obvious then that Magnetites #3 and #1 can perform well only at higher solids content (higher than 1.5 g/cm^3). For the same reasons Magnetite #1 cannot be used at high inlet pressures (Figure 10a) while the BMT35 bimodal magnetite will work at higher inlet pressures very well (Figure 10b). Comparison of the two data sets in Figure 12 also shows that, for the coarse feed particles (4 × 2 mm), lower Ep values were obtained with the use of more viscous (and more stable) Mag #2; on the other hand for the fine feed particles (0.5 × 0.355 mm) lower Ep values were obtained with the use of the less viscous Mag #1 medium. As our simulations of the DMC operation revealed, the shear rate of a 2 mm near density particle (density of the particle is 1.55 in the medium with density of 1.45 g/cm^3) inside a DMC cyclone with centrifugal acceleration of 40 m/s^2 remains slow in the medium with a high yield stress even at high density differential. These simulations also revealed that in the magnetites characterized by a high yields stress the size of near-density particles is critical; fine coal particles become "locked" in such a medium (He, Laskowski and Klein 2001). In practical terms, these effects manifest in lower sensitivity of coarse feed particles to medium rheology. These results were obtained at a fixed inlet pressure (10 × D) and will, of course, change at higher pressures.

One of the important questions which needs to be answered is whether current standards properly describe magnetite size distributions. Figure 6 also includes one experimental point at a dense medium of 1.55 relative density (taken from Figure 7). As seen, for this particular dense medium the separation efficiency was much better than for any other tested magnetite. This magnetite dense medium was a mixture of Mag #4 (very fine) and Mag #3 (very coarse), and Figure 7 shows further the effect of mixing ratios on separation efficiency (He and Laskowski 1995b). As this figure demonstrates, a bimodal particle size distribution is very beneficial.

Fine magnetites are characterized by high yield stress values. They are, therefore, quite stable even at low medium solids content (low medium density). With increasing medium density in such media the yield stress increases rapidly making separation difficult. For the same reason coarse magnetite should not be used at low medium density since dynamic stability for such a medium is very low. By increasing solids content the yield stress increases as well (Figure 4) and the system can be stabilized. Such magnetite media can be applied at higher medium densities (1.6–1.7). It is beneficial to apply higher inlet pressures when processing very fine coal. Because of poor dynamic stability of commercial magnetite media at high inlet pressures, finer magnetites work better under such conditions. The use of a bimodal magnetite brings about reduction of the yield stress with resulting improved separation efficiency.

For the reasons discussed above, different magnetites will work better over different medium density ranges. At low medium densities (<1.5 g/cm^3), a fine stable magnetite should be used. At high medium densities (>1.5 g/cm^3), it is better to use coarser magnetite to reduce the effect of medium rheology on DMC performance. While for coarser coal fractions the yield stress of the medium does not necessarily pose a serious problem and a stable magnetite medium may work very well, fine magnetite with a bimodal distribution will be superior for cleaning very fine coal (–0.5 +0.1 mm). The current industry standards entirely miss this point.

It is obvious that in practice it is crucial to maintain the optimum magnetite size distribution. At the Curragh mine this was achieved not only by more efficient magnetic recovery of fine magnetite, but also by the innovative transfer between coarse and fine coal DMC circuits to control magnetite size distribution (Kempnich, van Barneveld, and Luscan 1993). There is no doubt, however, that another important subject requiring further attention is development of a more efficient magnetic separation of fine magnetite particles (Klima and Killmeyer 1995).

SUMMARY

As this paper indicates, research in the area of fine coal cleaning in dense medium cyclones resulted in very significant achievements. The Greenside Colliery DMC fine coal cleaning plant in South Africa and the Curragh Mine DMC fine coal cleaning plant in Australia laid a foundation for the application of this new technology. As pointed out by Kempnich, van Barneveld and Luscan (1993), this was a result of progressive changes, not all in accordance with previously documented experience, to ultimately produce a successfully operating dense medium circuit for the beneficiation of fine coal. As often happens, these industrial applications were ahead of fundamental studies on the rheological properties of magnetite aqueous suspensions which only recently started catching up with what is happening in this area. It is quite likely that in the near future we will see more plants with various dense medium circuits which will also include processing of very fine coal. Further improvements in this technology may, however, be hindered if the fundamental studies such as those mentioned in this paper are not continued.

ACKNOWLEDGMENTS

Thanks are due to a number of people. First of all to my previous graduate students, Dr. Ying Bin He, Dr. Bern Klein, Dr. Qi Liu and Dr. Marek Pawlik, whose results have been heavily used in this paper.

REFERENCES

Barnes, H.A., J.F. Hutton, and K. Walters. 1989. *An Introduction to Rheology*. Amsterdam: Elsevier.

Berghofer, W. 1959. Konsistenz und Schwertrubeaufbereitung, *Bergbauwissenschaften,* 6: 493–504.

Collins, D.N., T. Turnbull, R. Wright, and W. Ngan. 1983. Separation efficiency in dense medium cyclones. *Trans. IMM, Sec. C.*, 92: 38–51.

Farris, R.J. 1968. Prediction of the viscosity of multimodal suspensions from unimodal viscosity data. *Trans. Soc. Rheol.*, 12: 281–300.

Firth, B.A., A.S. Swanson, and S.K. Nicol. 1978. The influence of feed size distribution on the staged flotation of poorly floating coals. *Proc. Australas. IMM*, No. 267, pp. 49–53.

Firth, B.A., A.S. Swanson, and S.A. Nicol. 1979. Flotation circuits for poorly floating coals. *Int. J. Min. Proc.*, 5: 321–334.

Geer, M.R. and M. Sokaski. 1963. Cleaning uni-sized fine coal in a DMC pilot plant. *U.S. Bureau of Mines Report,* No. 6274.

Harris, M.C. and J-P. Franzidis. 1995. A survey of fine coal treatment practice in South Africa. In: *Colloqium "Coal Processing, Utilisation and Control of Emissions,"* SAIMM, Mintek.

He, Y.B. and J.S. Laskowski. 1999. Rheological properties of magnetite suspensions. *Mineral Processing & Extractive Metal. Review*, 20: 167–182.

He, Y.B., J.S. Laskowski, and B. Klein. 2001. Particle movement in non-Newtonian slurries: the effect of yield stress on dense medium separation, *Chemical Eng. Sci.*, 56: 2991–2998.

He, Y.B. and J.S. Laskowski. 1994. Effect of Dense Medium Properties on the Separation Performance of a Dense Medium Cyclone. *Minerals Engineering*, Vol. 7: 209–222.

He, Y.B. and J.S. Laskowski. 1996. Separation of fine particles in dense medium cyclone: the effect of medium yield stress. In *New Trends in Coal Preparation Technologies and Equipment (Proc. 12th Int. Coal Prep. Congress, Cracow, 1994)*, S. Blaschke ed., New York: Gordon and Breach, pp. 175–184.

He, Y.B. and J.S. Laskowski. 1995. Dense medium cyclone separation of fine prticles. Part II. The effect of medium composition on DMC performance. *Coal Preparation*, 16: 27–49.

Kempnich, R., S. van Barneveld, and A. Luscan. 1993. Dense medium cyclones on fine coal— the Australian experience. *Proc. 6th Australian Coal Preparation Conference*, Mackay: Australian Coal Preparation Society, pp. 272–288.

Klassen, V.I. 1963. *Coal Flotation*, Moscow: Gosgortiekhizdat.

Klein, B., S.J. Partridge, and J.S. Laskowski. 1990. Rheology of unstable mineral suspensions. *Coal Preparation*, 8: 123–134.

Klein, B., J.S. Laskowski, and S.J. Partridge. 1995. A new viscometer for rheological measurements on settling suspensions. *J. of Rheology*, 39: 827–840.

Klima, M.S., R.P. Killmeyer, and R.E. Hucko. 1990. Development of a micronized magnetite cycloning process. In: *Proc. 11th Int. Coal Prep. Congress*, Tokyo: Min. & Materials Processing Inst. of Japan, Tokyo, pp. 145–150.

Klima, M.S. and R.P. Killmeyer. 1995. An evaluation of a laboratory wet-drum magnetic separator for micronized magnetite recovery. *Coal Preparation*, 16: 203–215.

Mikhail, M.W. and D.G. Osborne. 1990. Magnetite heavy media: standards and testing procedures. *Coal Preparation*, 8: 111–121.

O'Brien, E.J. and K.J. Sharpeta. 1978. Water-only cyclones: their function and performance. In: *Coal Age Operating Handbook of Preparation*, Edited by P. Merit. New York: McGraw-Hill, pp. 114–118.

Robertson, A.C. and J.D. Williams. 1991. Magnetite production for the coal industry. In: *Queensland Coal Symposium*, Brisbane: Australia. IMM, pp. 83–87.

Stoessner, R.D., D.G. Chedgy, and E.A. Zawadzki. 1988. Heavy medium cyclone cleaning 28 × 100 mesh raw coal. In: *Industrial Practice of Fine Coal Processing*, Edited by R.R. Klimpel and P.T. Luckie. Littleton: SME, pp. 57–64.

van der Walt, P.J, L.M. Falcon, and P.J.F. Fourie. 1981. Dense medium separation of minus 0.5 mm coal fines. *Proc. 1st Australian Coal Preparation Conference*, Newcastle: Australian Coal Preparation Society, pp. 207–219.

Methodology for Performance Characterization of Gravity Concentrators

M. Nombe[*], J. Yingling[*], and R. Honaker[*]

Mathematical models of partition curves are routinely used to simulate the performance of density-based separations. These models are two-dimensional in that they represent the probability of a particle reporting to the product stream as a function of the relative particle density. However, the particle size distribution of the feed significantly affects performance, which is commonly quantified by additional mathematical functions. In this publication, the practical fitting and use of a mathematical modeling approach is discussed which employs a single functional form that simultaneously accounts for the effects of particle density and size. This approach allows the generation of a partition surface rather than a partition curve for a given set of operating conditions in a density separator. The three-dimensional model provides a more accurate prediction of separation performance by better accounting for size effects. Moreover, the model can be generalized similar to standard partition curves, making it easy to model the effect of changes in control settings on separator performance.

INTRODUCTION

The performance of density-based separators is commonly assessed and predicted using Tromp or partition curves that describe the probability of a particle to report to the product stream as a function of the particle density (Tromp, 1937). In the absence of experimental data, mathematical models are used to predict the partition values. These models vary significantly in form including normal, log-normal, logistic, Weibull and various exponentials (Peng and Luckie, 1991). The model providing the best fit to the performance data is dependent on the separator type. However, recent efforts have addressed the potential of using an overall generalized model that accurately describes the performance behavior of most unit operations as a function of particle density (Kelly and Subasinghe, 1991; Klima and Luckie, 1997).

[*] Dept. of Mining Engineering, University of Kentucky, Lexington, Ky.

Nearly all of the performance models predict partition curves as a function of the particle density. However, it is well known that the performance varies significantly as a function of particle size. To address the performance changes, efficiency parameter values associated with each particle size fraction in the feed need to be considered separately and individual partition curves for each size fraction determined. As such, the family of performance curves generated from the multiple determinations represents the broad particle size ranges within the feed stream. A more accurate approach would involve the use of a single model to obtain a three-dimensional performance evaluation as a function of both particle density and size.

A NEW APPROACH FOR FITTING GENERALIZED DISTRIBUTION CURVES

Yingling et al. (2002) have developed a mathematical model that has a logit form and will provide the aforementioned three-dimensional performance assessment. The model considers a unit process having a performance represented by the partition number $P(x,\rho)$, i.e.,

$$P(x, \rho) = \frac{e^{h(x, \rho)}}{1 + e^{h(x, \rho)}} \qquad \text{(EQ 1)}$$

in which $h(x,\rho)$ is a function describing the performance efficiency as a function particle size, x, and particle density, ρ. The functional relationship describing $h(x,\rho)$ is linear in some cases and thus takes the form:

$$h(x, \rho) = \alpha + \beta_1 x + \beta_2 \rho \qquad \text{(EQ 2)}$$

in which α, β_1 and β_2 are estimated from the washability data for the feed and product stream.

However, interactions between particle size and density in predicting the logit values may require the use of a bilinear model, i.e.,

$$h(x, \rho) = \alpha + \beta_1 x + \beta_2 \rho + \beta_{12} x\rho \qquad \text{(EQ 3)}$$

To determine the appropriate model for $h(x,\rho)$, one simply computes and then plots what are called the logit transforms of the experimentally determined partition values, $P(x,\rho)$. The logit transform, $g(x,\rho)$ is computed for each size and density interval using the formula:

$$g(x, \rho) = \ln\left(\frac{P(x, \rho)}{1 - P(x, \rho)}\right) \qquad \text{(EQ 4)}$$

The logit transform values for a particular size x are plotted versus relative density and a "best-fit" line (one can simply "eyeball" the line) is drawn on the graph. Using another set of data for a different size fraction, the process is repeated. If the slopes of the lines generated from Equation 2 are equal, than the interactive effect is small and the simple linear model can be used. The interactive parameter effect is considered significant when the line slopes are not equal, for which case the bilinear model should be employed. In our experience in fitting models to various vessels, sometimes it is necessary to transform the size values by taking the natural log of those values so that the logit lines have a linear form. Once the form of the model is known, various mathematical approaches, including regression analysis, might be used for fitting the parameters. Yingling et al. (2002) present an approach using the statistical technique of maximum likelihood estimation that has advantages relative to standard regression approaches in accounting for the effects of the feed mass distribution in fitting the model's parameters.

So far, model fitting has been described using a single experimental data set. Yingling et al. (2002) have determined that these partition surfaces can be generalized similar to the way traditional partition curves can be generalized (e.g., Gottfried, 1978). That is, a single generalized curve can model separator performance from different experiments conducted under variable conditions. To do this, a reference size fraction is selected, preferably the middle particle size fraction. Then, the d_{50} value is fit for that size fraction for all experiments. One experiment is picked as a reference case and referred to as experiment p. Then, the parameters for the bilinear model for all of the various other cases s will be related to the parameters in case p by the equations:

$$\alpha^{<q>} = \alpha^{<p>}$$

$$\beta_1^{<q>} = \beta_1^{<p>}$$

$$\beta_2^{<q>} = \frac{d_{50}^p}{d_{50}^q}\beta_2^{<p>}$$

$$\beta_{12}^{<q>} = \frac{d_{50}^p}{d_{50}^q}\beta_{12}^{<p>}$$

(EQ 5)

(Note: If we use a linear model form, the last equation above is not needed.) Hence, only the knowledge of how d_{50} for the reference size fraction changes from experiment to experiment is needed in order to adjust the parameters of the model. Thus, a single, simple model is derived that shows how separation performance changes as control settings and operating conditions for the machine vary. Yingling et al. (2002) have given a maximum likelihood based procedure that can combine data from multiple experiments and best-fit a single generalized distribution surface.

To summarize, the methodology involved in developing the 3-D partition surface as a function of particle density and size is schematically illustrated in Figure 1. The logit model can be used to produce partition number predictions over an unlimited range of particle size and density fractions between the outer limits of the experimental data used to determine the initial constant values.

PERFORMANCE PREDICTION APPLICATIONS

Case Study on Dense Medium Vessels Data

Hudy (1968) conducted six experiments on dense medium coarse-coal vessels. The feed washability data, the product washability data, the size and densities were used to determine α_1, β_1, β_2, and β_{12}. The estimated values of the logit transform parameters are in Table 1.

The linearity of the curves in Figure 2 suggested that independent effect of particle size is small and interaction between particle size and density is significant. Since the curves are parallel straight lines and the slopes differ by size fraction, a bilinear form (Equation 3) was used to fit the model to the experimental data. Figures 3–5 show the fit of the model to the experimental data. When examining all three selected cases, a good

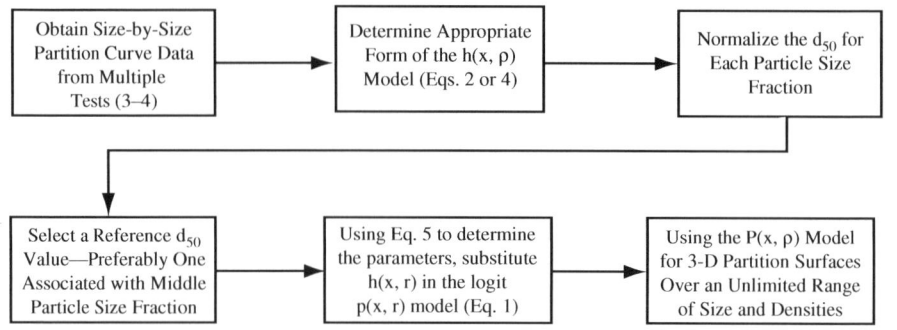

FIGURE 1 A schematic of the 3-D partition surface modeling as a function of particle density and size

TABLE 1 The computed parameters for the data set from Hudy (1968)

EXP	α	β_1	β_2	β_{12}
GPC	36.075	−3.4454	−71.3	6.82
1	36.075	−3.4454	−24.290	2.3270
2	36.075	−3.4454	−24.869	2.3824
3	36.075	−3.4454	−26.554	2.5439
4	36.075	−3.4454	−22.615	2.1666
5	36.075	−3.4454	−26.309	2.5204
6	36.075	−3.4454	−20.107	1.9263

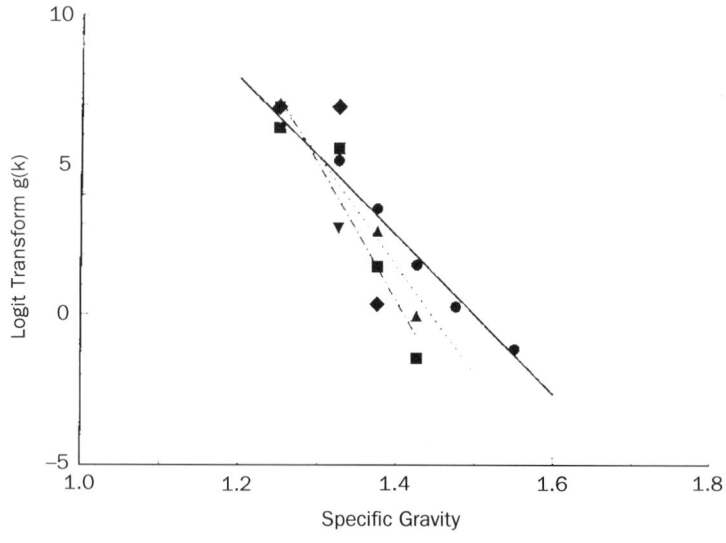

FIGURE 2 Logit transform vs specific gravity for various size fraction (experiment 1 from Hudy 1968)

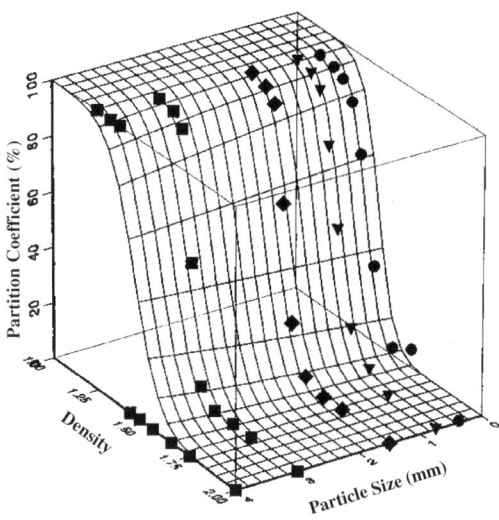

FIGURE 3 3-D model partition surface fitted to set 1 experimental data of Hudy (1968)

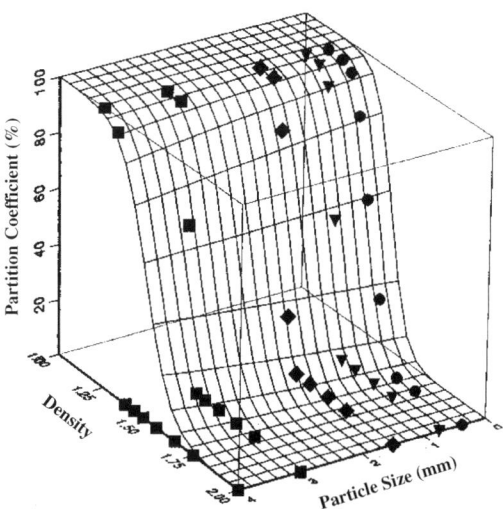

FIGURE 4 3-D model partition surface fitted to set 3 experimental data of Hudy (1968)

fit is realized for each case. The independent effect of particle size is small since the surface does not shift significantly as the particle size changes.

Case Study on Dense Medium Cyclones Data

Duerbrouck and Hudy (1972) evaluated eight dense-medium cyclone vessels. As in the previous case study, the feed and product washability data as well as the particle size and density values were used to determine α_1, β_1, β_2, and β_{12}. The estimated values of the logit transform parameters are presented in Table 2. It can be seen in Figure 6 that

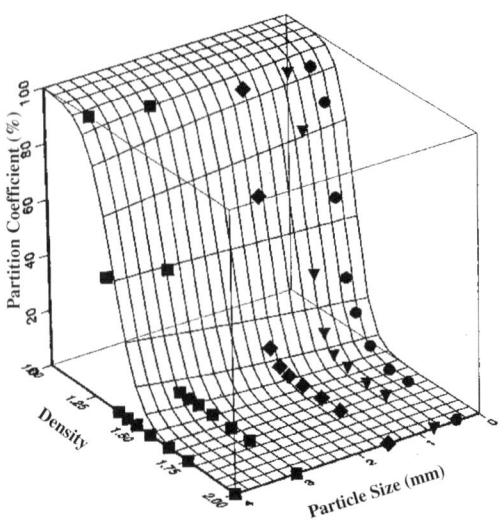

FIGURE 5 3-D model partition surface fitted to set 2 experimental data of Hudy (1968)

TABLE 2 The computed parameters for data set from Duerbrouck and Hudy (1972)

EXP	α	β_1	β_2	β_{12}
GPC	59.781	−6.2786	−59.86	6.55
1	59.781	−6.2786	−38.807	4.2467
2	59.781	−6.2786	−39.462	4.3183
3	59.781	−6.2786	−39.012	4.2691
4	59.781	−6.2786	−42.007	4.5968
5	59.781	−6.2786	−40.416	4.4227
6	59.781	−6.2786	−41.896	4.5849
7	59.781	−6.2786	−36.421	3.9855
8	59.781	−6.2786	−41.654	4.5581

all curves except for the large particles follow a linear trend. The linearity of the lines suggest that the independent effect of particle size on the partition curves is small and the interaction between particle size and relative density is significant. The bilinear model (Equation 3) was used to fit the model surface to the experimental data as shown in Figures 7 and 8. One can observe that the fit is good and the shapes of curves do not shift significantly as particle size changes.

SUMMARY AND CONCLUSIONS

It has been demonstrated that the concept of generalized partition curves can be extended from two-dimensional curves that give separator performance by density to three-dimensional surfaces that give separator performance by both particle density and size. In contrast to other approaches that have been used to model partition curves as a function of both particle size and density, the bilinear logistic models are parsimonious in

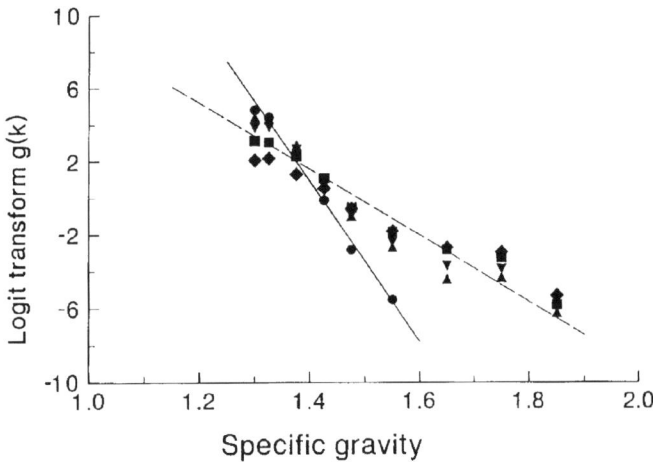

FIGURE 6 Logit transform vs specific gravity for various size fraction (experiment 1 from Duerbrouck and Hudy 1972)

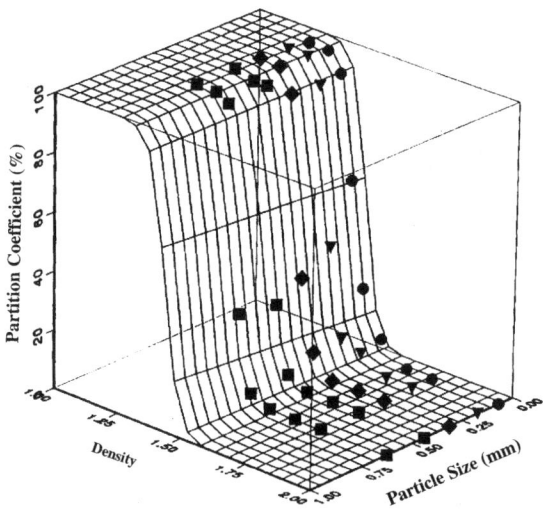

FIGURE 7 3-D model partition surface fitted to set 6 experimental data of Duerbrouck and Hudy (1972)

the use of parameters, requiring only four. Moreover, the concept of generalized partition curves, as discussed in this paper, simplifies the task of describing changes in separation behavior as a function of changes in vessel's operating conditions and design/control variables. The flexibility of the model to fit a wide range of performances achieved by separation vessels has been increased using this approach by considering the independent effect of particle size and the interaction between particle size and relative density. In addition, the model can more accurately characterize separator performance as a function of particle size by using a surface that represents all sizes rather than the more common approach of using a series of curves that typically only model a few discrete size ranges.

78 | GRAVITY CONCENTRATION FUNDAMENTALS

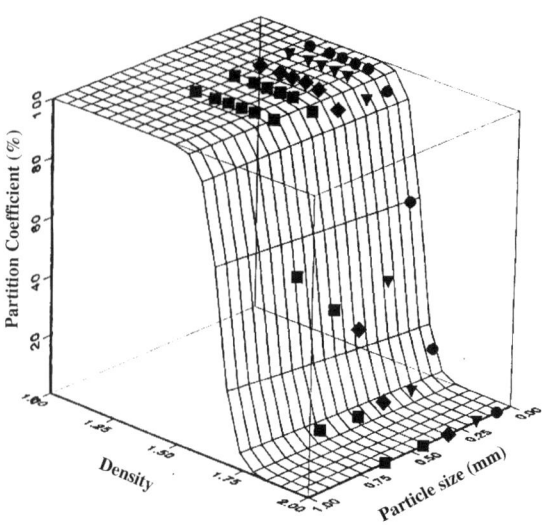

FIGURE 8 3-D model partition surface fitted to set 7 experimental data of Duerbrouck and Hudy (1972)

Finally, the three-dimensional models show good fit to experimental data, indicating the robustness of the approach.

REFERENCES

Duerbrouck, A.W., and J. Hudy. 1972. Performance characteristics of coal-washing equipment: dense medium cyclones. U.S. Bureau of Mines. RI 7673.

Gottfired, B.S. and P.S. Jacobsen. 1978. Generalized distribution curve for characterizing the performance of coal-cleaning equipment. U.S. Bureau of Mines. RI 8238.

Hudy, J. Jr. 1968. Performance characteristics of coal-washing equipment: dense-medium coarse-coal vessels. U.S. Bureau of Mines. RI 7154.

Kelly, E.G. and G.K.N.S. Subasinghe. 1991. Gravity performance curves: a re-examination. *Minerals Engineering*. 4. 1207–1218.

Klima, M. and P.T. Luckie. 1997. Towards a general distribution model for flowsheet design. *Coal Preparation*. 20. 3–11.

Tromp, K.F. 1937. New methods of computing the washabilities of coals. (a) *Glukauf*, 37, 125–156; (b) *Colliery Guardian*. 154. 955–959. 1009.

Yingling, J., X.Y. Wang, X.H. Wang, and R.Q. Honaker. 2002. Fitting and application of generalized partition curves as predictive models for gravity separation vessels. Accepted for publication in *International Journal of Mineral Processing*.

SECTION 2

Coal-based Gravity Separations

- Optimum Cutpoints for Heavy Medium Separations   **81**
- Operating Characteristics of Water-only Cyclone/Spiral Circuits Cleaning Fine Coal   **93**
- Comparing a Two-stage Spiral to Two Stages of Spirals for Fine Coal Preparation   **107**
- Advances in Teeter-bed Technology for Coal Cleaning Applications   **115**
- Innovations in Fine Coal Density Separations   **125**

Optimum Cutpoints for Heavy Medium Separations

G.H. Luttrell[*], C.J. Barbee[*], and F.L. Stanley[†]

Heavy media baths and cyclones are commonly used to remove unwanted impurities from coarser fractions of run-of-mine coal. Heavy media circuits presently account for approximately half of the installed plant capacity in the United States and are responsible for the production of nearly 250 million tons of clean coal annually. Unfortunately, significant amounts of recoverable clean coal are often lost in heavy medium circuits due to problems associated with the selection of specific gravity setpoints. This article reviews the underlying problems associated with the optimization of heavy medium circuits and, based on this analysis, provides recommendations for circuit operation that can be adopted to maximize clean coal yield.

INTRODUCTION

The flowsheet for a coal preparation plant can typically be represented by a series of sequential unit operations for sizing, cleaning, and dewatering (see Figure 1). This sequence of processing steps is repeated several times since the processes employed in modern plants have a limited range of applicability in terms of particle size. As a result, modern plants may include as many as four separate processing circuits for treating the coarse (plus 50 mm), intermediate (50 × 1 mm), fine (1 × 0.15 mm), and ultrafine (minus 0.15 mm) material. The clean coal products from these circuits are typically blended back together prior to shipment to market.

Most of the coarse coal circuits in modern coal preparation plants make use of heavy medium processes such as baths and cyclones. During production runs, appropriate specific gravity cutpoints must be selected for these operations to ensure that the overall clean coal product from the plant meets the quality criteria dictated by the sales contracts. Plant operators will often select cutpoints that produce the same clean coal quality in every circuit throughout the plant. For example, the heavy medium bath and heavy

---

[*] Dept. of Mining and Minerals Engineering, Virginia Polytechnic Institute and State University, Blacksburg, Va.

[†] Pittston Coal Management Co., Lebanon, Va.

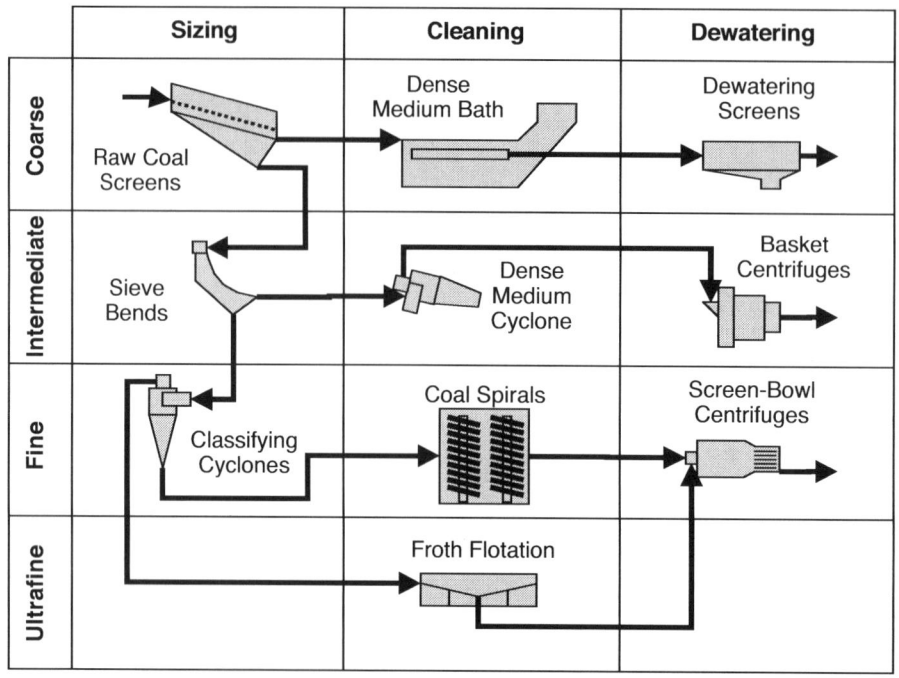

FIGURE 1 Flowsheet matrix for a modern coal preparation plant incorporating four parallel circuits

medium cyclones in a given plant could all be set to produce the same clean coal ash of 10% in order to satisfy a 10% ash contract. While this approach guarantees that the target quality is met, this method of selecting cutpoints may not provide the highest overall yield of clean coal. It may be possible to obtain a higher yield by operating the heavy medium bath at 12% ash and the heavy medium cyclones at 9% ash in order to obtain a final blended product containing 10% ash. Although there is no difference in the quality of the product shipped to the customer, the net increase in plant yield obtained by optimizing the cutpoints of the heavy medium circuits will often have a tremendous impact on plant profitability. The underlying principles of this important optimization concept are discussed in greater detail in the following sections.

OPTIMIZATION OF PLANT YIELD

Incremental Quality Concept

The clean coal yield (Y) and quality (Q) for a plant consisting of N parallel circuits can be calculated from simple weighted averages using:

$$Y = \sum_{i=1}^{N} M_i Y_i \quad \text{(EQ 1)}$$

$$Q = \frac{\sum_{i=1}^{N} M_i Y_i Q_i}{\sum_{i=1}^{N} M_i Y_i} \quad \text{(EQ 2)}$$

in which M_i is the fractional mass of feed coal reporting to circuit i, Y_i is the clean coal yield from the separator in circuit i, and Q_i is the coal quality produced by the separator in circuit i. In most cases, the total plant yield (Y) is limited by one or more constraints imposed on total product quality (Q).

Many plant operators select cutpoints that ensure that the quality constraints are not exceeded in any given circuit, i.e., all circuits are set to produce the same quality. However, the optimum operating points for different circuits are those that maximize overall plant yield at a given clean coal quality. Depending on the liberation characteristics of the feed coal, this may or may not require identical product qualities for each circuit. A common method used to identify these optimum cutpoints is to conduct mathematical simulations using experimental float-sink data for different size fractions of the feed coal and empirical partition curves for the different separation processes. The simulations are repeated for all possible cutpoints for each circuit and the combination that provides the highest plant yield at the desired target quality is selected as the optimum (Peng and Luckie, 1991). Unfortunately, this trial-and-error approach provides little insight regarding the underlying principles that influence the optimization process. As a result, a more attractive method for plant optimization is to use the concept of *constant incremental quality*. This concept, which has long been recognized in coal preparation (Mayer, 1950; Dell, 1956; Abbott, 1982; Rayner, 1987), states that the clean coal yield for parallel operations is maximum when all circuits are operated at the same incremental quality. This statement is true for any number of parallel circuits and is independent of the particle size and washability characteristics of the feed coal.

The mathematical proof of the incremental quality concept is rather simple for a two-circuit plant. According to Equation 1, the combined yield obtained from two circuits operating in parallel can be determined from:

$$Y = M_1 Y_1 + M_2 Y_2 \quad \text{(EQ 3)}$$

To maximize the value of Y, Equation 3 must be differentiated with respect to Y_1 and set equal to zero. This calculation gives:

$$\frac{\partial Y}{\partial Y_1} = M_1 + M_2 \frac{\partial Y_2}{\partial Y_1} = 0 \quad \text{or} \quad \frac{\partial Y_2}{\partial Y_1} = -\frac{M_1}{M_2} \quad \text{(EQ 4)}$$

Likewise, Equation 2 can be rearranged and used to provide a second governing expression for Y, i.e.:

$$Y = \frac{M_1 Y_1 Q_1 + M_2 Y_2 Q_2}{Q} \quad \text{(EQ 5)}$$

As with Equation 3, Equation 5 can also be maximized by taking the derivative of Y with respect to Y_2 and setting the result equal to zero. This calculation gives:

$$\frac{\partial Y}{\partial Y_2} = \frac{M_1}{Q}\left(Y_1\frac{\partial Q_1}{\partial Y_2} + Q_1\frac{\partial Y_1}{\partial Y_2}\right) + \frac{M_2}{Q}\left(Y_2\frac{\partial Q_2}{\partial Y_2} + Q_2\right) = 0 \qquad \text{(EQ 6)}$$

$$-\frac{M_1}{M_2}\left(Y_1\frac{\partial Q_1}{\partial Y_2} + Q_1\frac{\partial Y_1}{\partial Y_2}\right) = Y_2\frac{\partial Q_2}{\partial Y_2} + Q_2 \qquad \text{(EQ 7)}$$

Finally, substitution of Equation 4 into Equation 7 provides the final expression for this problem, i.e.:

$$Y_1\frac{\partial Q_1}{\partial Y_1} + Q_1 = Y_2\frac{\partial Q_2}{\partial Y_2} + Q_2 \qquad \text{(EQ 8)}$$

Although not entirely obvious at first inspection, the terms on each side of Equation 8 are mathematically equivalent to the incremental quality in circuits 1 and 2, respectively. Incremental quality is simply the effective quality of the last material added to the clean coal (or removed from the refuse) when the yield is increased by an infinitesimal amount. Equation 8 provides the mathematical proof that each circuit should be operated at the same incremental quality in order to maximize yield.

Correlation with Specific Gravity

Incremental quality is a mathematical term that cannot be directly monitored in most industrial systems. However, this value can be estimated for ideal separations if the quality parameter of interest is ash. This approximation is based on the assumption that run-of-mine coals contain only two components, i.e., a low density, ash free carbonaceous component and a high density, pure ash mineral component. Based on this assumption, the ash content of a given particle must increase linearly with the reciprocal of particle density (ρ) according to the expression:

$$\text{Ash}(\%) = \frac{100}{\rho_2 - \rho_1}\left[\frac{\rho_2\rho_1}{\rho} + \rho_2\right] \qquad \text{(EQ 9)}$$

where ρ_1 and ρ_2 is the density of the light (carbonaceous) and dense (mineral) components, respectively (Anon., 1966). A detailed analysis of this problem has been presented by Abbot and Miles (1990).

The utility of Equation 9 can be demonstrated by plotting the ash contents of narrowly partitioned density fractions of coal obtained from standard float-sink tests. For example, these values are plotted in Figure 2 for six different size fractions of a run-of-mine coal. For particles coarser than 28 mesh, the data show that the same incremental ash is obtained for a given specific gravity regardless of the size fraction treated. The deviation noted for the fractions finer than 28 mesh can normally be attributed to inefficiencies in the experimental float-sink procedures. Data collected to date suggest that slightly different linear relationships may be obtained for bituminous coals of different rank (i.e., high-, medium- and low-volatile matter contents) due to the density variations in the carbonaceous matter. The impacts of the variations must be evaluated on a case-by-case basis. In addition, the presence of a third density component (e.g., pyrite) in some coal fractions has also been known to create minor deviations from the linear relationship dictated by Equation 9. In any case, the incremental quality concept states that the clean coal yield will be maximized for a plant constrained by an upper limit on

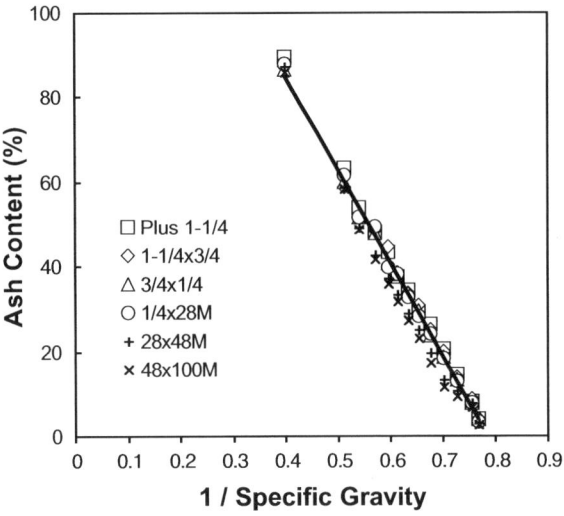

FIGURE 2 Relationship between specific gravity and particle ash content for different size fractions of run-of-mine coal

ash when all parallel circuits are operated at the same incremental ash. Since Equation 9 shows that incremental ash is fixed by specific gravity, this optimization concept can now be extended to state that plant performance can be optimized by operating all circuits at the same specific gravity cutpoint (Clarkson, 1992). This statement is true regardless of the size distribution or washability characteristics of the feed coal, provided that ideal (or very efficient) separations are maintained in each circuit.

Effects of Misplaced Material

Plant optimization would be relatively straightforward were it not for the presence of misplaced particles. A plant equipped with ideal separators could achieve optimum performance simply by maintaining the same specific gravity cutpoint in parallel circuits. Unfortunately, this approach must be modified since ideal separators do not exist in practice. The impact of inefficiencies on the selection of optimum circuit cutpoints can be studied using a variety of mathematical techniques. The most common approach is to convert the "ideal" separation curves obtained from standard characterization tests (e.g., float-sink tests) into "actual" separation curves. This conversion can be accomplished using empirical partition models (Armstrong and Whitmore, 1982; Rong and Lyman, 1985). The optimum cutpoints can then be identified using one of several plant simulation programs that are now commercially available for this purpose. In addition, many of the advanced spreadsheet programs equipped with built-in optimization routines make such calculations possible for all but the most complicated plant circuits.

Figure 3 shows the results of partition simulations conducted using a hypothetical set of float-sink data. In this case, the process efficiency was varied from very good (Ep = 0.02) to very poor (Ep = 0.16) while the specific gravity cutpoint (SG_{50}) was held constant at 1.5 SG. The partition curves corresponding to these conditions are shown in Figures 3a and 3b, respectively. These curves represent the fraction of material in a given specific gravity class that is present in the feed that reports to clean coal. When using the sharp partition curve (Ep = 0.02), all of the low-density (high quality) material present in the feed coal

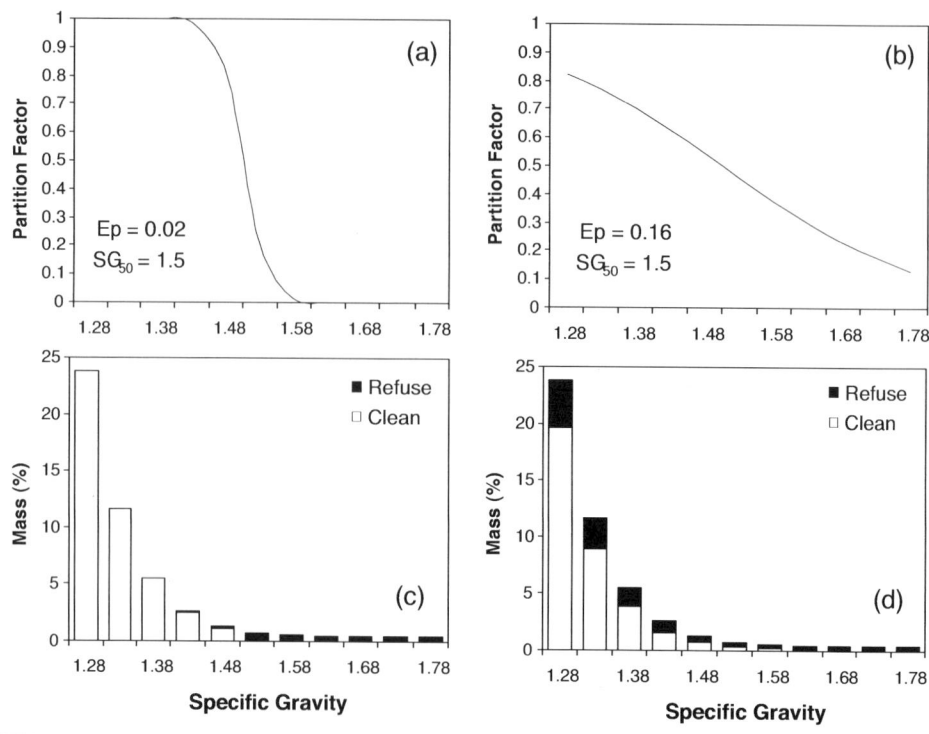

FIGURE 3 Effect of separator efficiency (Ep) on the misplacement of material in different SG classes

reported to the clean coal product (Figure 3c). In contrast, Figure 3d shows that the poor partition curve (Ep = 0.16) resulted in significant misplacement of low-density material to the refuse stream. The misplaced material has the effect of lowering the incremental ash of a product generated at a particular specific gravity. This is because the mass present in typical run-of-mine coals increases with decreasing specific gravity in the region where most industrial separations occur (i.e., below ≈1.7 SG). Consequently, a greater proportion of lower density middlings is misplaced into the refuse stream than higher density middlings into the clean coal stream. The shift of higher quality (lower ash) material lowers the effective incremental ash. Therefore, less efficient circuits must be operated at a higher specific gravity cutpoint in order to maintain the same incremental quality (Clarkson, 1992). In general, efficiencies of density-based separators tend to decline as the particle size decreases. Consequently, circuits treating finer particles must typically use correspondingly higher specific gravity cutpoints to maintain optimum yield.

To further illustrate the effect of misplaced material on incremental quality, several additional series of simulation runs were conducted as a function of SG_{50} and Ep (see Figure 4). Incremental ash was determined for each simulation run for SG_{50} values ranging from 1.3 to 1.8 SG and for five different Ep values ranging from 0.01 to 0.16. In order to represent feed coals with different cleanability characteristics (i.e., different washabilities), the percentage of middlings material present in the hypothetical float-sink data were adjusted over a wide range of values. To quantify these changes, a cleanability index was established to serve as a relative indicator of the potential cleanability of each type of feed coal. The cleanability index (CI), which can vary from 0 to 1, is mathematically defined as:

OPTIMUM CUTPOINTS FOR HEAVY MEDIUM SEPARATIONS | 87

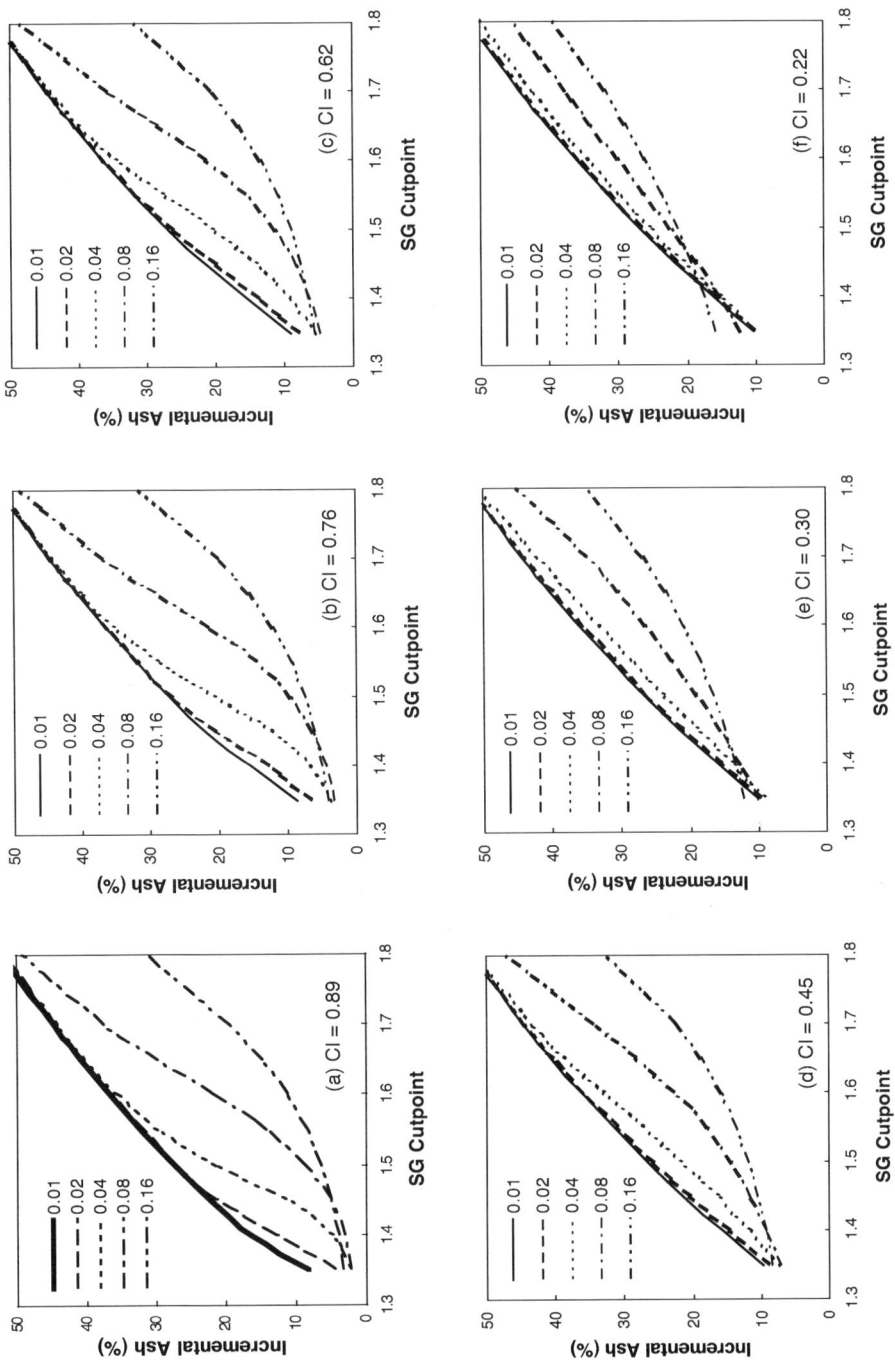

FIGURE 4 Effect of SG_{50} and Ep on the incremental ash for six different feed coals (a—best washability, f—worst washability)

$$CI = \frac{Y_{1.3}}{Y_{1.6}} \quad \text{(EQ 10)}$$

in which $Y_{1.3}$ and $Y_{1.6}$ are the clean coal yields obtained from the washability data at specific gravities of 1.3 and 1.6 SG, respectively. Obviously, a feed coal that responds well to density separations would have a very high CI value, while a poorly responding feed coal would have a low CI value. Simulations were conducted for CI values ranging from a high of 0.89 (Figure 4a) to a low of 0.22 (Figure 4f).

Several important conclusions may be derived from the data shown in Figure 4. First, the data show that the incremental ash increased in all cases as the specific gravity cutpoint was raised from 1.3 to 1.8 SG. This trend should be expected in light of the preceding discussions describing the association between specific gravity and incremental quality. Second, the same incremental ash versus specific gravity curve was obtained for all coal types (i.e., for all CI values) when Ep values of 0.02 or less were employed. This shows that efficient separators should be operated at the same specific gravity cutpoint to maximize yield irrespective of the washability characteristics of the feed coal and the percentages of mass reporting to each circuit. Third, the simulation data show that less efficient separations (Ep > 0.08) require a significantly higher specific gravity cutpoint in order to maintain the same incremental quality. The magnitude of the increase varies in accordance with the washability characteristics of the feed coal. Coals that are generally easier to clean (higher CI) require larger increases in SG cutpoint for a given Ep than coals that are more difficult to clean (lower CI). This difference is due to the increased steepness of the yield versus specific gravity curve for higher quality coals in the specific gravity ranges where industrial separations typically occur. This finding indicates that easy to clean coals are more sensitive to changes in Ep than are difficult to clean coals.

Approximate Corrections for Misplaced Material

Detailed simulations are required in order to accurately determine the specific gravity cutpoints required to optimize separations involving poor Ep values. Unfortunately, this mathematical exercise can be time consuming and require large amounts of experimental data that may not be readily available. In light of these difficulties, a variety of numerical simulations were conducted to establish "approximate" corrections to specific gravity cutpoints as a function of separation efficiency (i.e., Ep) and coal type (i.e., CI). To simplify this procedure, coal type was broadly classified into four general categories of relative cleanability, i.e., easy (CI > 0.50), moderate (0.35 < CI < 0.50), difficult (0.25 < CI < 0.35), and very difficult (CI < 0.25). For each coal type, approximate corrections were determined for different specific gravity ranges and Ep values. These corrections, which are summarized in Table 1, indicate in absolute points how much the specific gravity must be increased in order to maintain the same incremental ash that would be obtained using an ideal separator at a given specific gravity cutpoint. For example, Figure 4d shows that a cutpoint of 1.54 SG would be required to provide an incremental ash of 30% for an efficient separator (Ep < 0.02) and moderate coal cleanability (CI = 0.45). Likewise, this plot shows that a separator characterized by an Ep of 0.16 would need to operate at a higher cutpoint of 1.77 SG to keep the same incremental ash of 30%. This represents an increase in cutpoint of 0.23 SG units (i.e., 1.77 – 1.54 = 0.23). In the absence of this simulation data, this absolute difference can be estimated directly from the values provided in Table 1. This chart also shows that a separation characterized by a poor efficiency (Ep = 0.16) and moderate cleanability (CI = 0.45) would need to operate at 0.23 SG points higher than an efficient separator to maintain the same incremental

TABLE 1 Approximate increase in SG cutpoint for various coal types and efficiency (Ep) values

Cleanability	SG Range	Ep = 0.04	Ep = 0.08	Ep = 0.16
Easy (CI > 0.50)	<1.4	0.04	0.14	0.23
	1.4–1.5	0.04	0.14	0.24
	1.5–1.6	0.04	0.13	0.25
	>1.6	0.02	0.09	0.22
Moderate (0.35 < CI < 0.50)	<1.4	0.03	0.09	0.16
	1.4–1.5	0.03	0.11	0.19
	1.5–1.6	0.03	0.12	0.23
	>1.6	0.03	0.10	0.22
Difficult (0.25 < CI < 0.35)	<1.4	0.02	0.05	0.09
	1.4–1.5	0.02	0.08	0.14
	1.5–1.6	0.03	0.11	0.20
	>1.6	0.03	0.10	0.21
Very Difficult (C < 0.25)	<1.4	0.01	0.00	0.00
	1.4–1.5	0.01	0.02	0.03
	1.5–1.6	0.01	0.05	0.08
	>1.6	0.01	0.05	0.10

ash. This correction chart may be useful in cases where full numerical simulations to determine optimum cutpoints are impractical.

CASE STUDIES

The usefulness of the optimization approach described above was evaluated by comparing the optimum SG values obtained for two run-of-mine coals versus those predicted using Table 1. The first coal (Coal A) was selected because it was known to be easy to upgrade. This coal provided a high clean coal yield and low ash content even at a relatively high cutpoint. In contrast, the second coal sample (Coal B) was selected because it was known to contain large amounts of middlings particles and, as a result, was relatively difficult to treat by heavy medium processes. The CI values for Coal A and Coal B were determined to be 0.77 and 0.27, respectively.

Detailed mathematical simulations were conducted and the optimum cutpoints were determined for Ep values of 0.02 and 0.08 for each of the two coals. These optimum cutpoints were then compared with those obtained by the estimation procedure described previously. The results of these comparisons are summarized in Table 2. In general, the estimated values compare very favorably with the actual cutpoints obtained by the very laborious simulation method. In all cases, the exact simulated values and the estimated values differed by less than 0.02 SG units. Although not shown in this manuscript, very good predictions were also obtained for a variety of coal samples representing a wide range of coal types. Nevertheless, the estimation procedure should be used only as a rough guide for estimating cutpoints in cases where full simulations to accurately establish the optimum cutpoints cannot be performed. Significant errors may occur as a result of unexpected variations in feed coal washability or separator characteristics that may make these predictions less reliable.

TABLE 2 Optimum SG cutpoints obtained by estimation and detailed simulation

Coal Type	Target SG_{50} (Ep = 0.00)	Estimated* Correction @ Ep = 0.08	Estimated SG_{50} @ Ep = 0.08	Simulation SG_{50} @ Ep = 0.08
Coal A (CI = 0.77)	1.40	0.14	1.54	1.52
	1.50	0.13	1.63	1.62
	1.60	0.09	1.69	1.70
Coal B (CI = 0.27)	1.40	0.02	1.42	1.43
	1.50	0.05	1.55	1.56
	1.60	0.05	1.65	1.67

* Corrections obtained from Table 1.

CONCLUSIONS

The performance of heavy medium circuits can be optimized by selecting optimum cutpoints that maximize product yield at a given clean coal quality. These optimum cutpoints may be readily identified using a theoretical principle known as the incremental quality concept. According to this concept, a plant limited by an upper constraint on clean coal quality will produce maximum total yield when all parallel circuits are operated at the same incremental quality. For plants limited by product ash content, this requirement is met when efficient circuits are operated at the same specific gravity cutpoints. Less efficient circuits, on the other hand, need to be operated at slightly higher cutpoints to correct for the effects of misplaced coal and middlings on incremental ash content. The magnitude of the increase in cutpoint required to optimize performance can be determined using mathematical simulations involving empirical partition curves. Unfortunately, this laborious technique is time consuming and may be impractical in cases where insufficient data is available to fully characterize the washability of the feed coal. To help alleviate this problem, an estimation procedure has been developed that is capable of roughly predicting the optimum cutpoints that are needed in less efficient circuits. Only two parameters are required to use this procedure, i.e., the relative efficiency (Ep) of the process and the cleanability index (CI) of the feed coal. The cleanability index is a dimensionless term that provides an indication of the relative mass proportions of pure coal and middlings present in a given feed coal. This value can be calculated from float-sink tests conducted at just two SG values (i.e., 1.3 and 1.6). Case studies conducted with two different run-of-mine coals suggest that the proposed estimation procedure provides good estimates of the specific gravity cutpoints that are required to optimize the performance of heavy medium processes.

REFERENCES

Abbott, J., 1982. The Optimisation of Process Parameters to Maximise the Profitability from a Three-Component Blend, 1st Australian Coal Preparation Conf., April 6–10, Newcastle, Australia, 87–105.

Abbott, J. and Miles, N.J., 1990. Smoothing and Interpolation of Float-Sink Data for Coals, Inter. Symp. on Gravity Separation, Sept. 12–14, Cornwall, England.

Anonymous, 1966. Plotting Instantaneous Ash Versus Density, *Coal Preparation*, Jan.-Feb., 2(1): 35.

Armstrong, M. and Whitmore, R.L., 1982. The Mathematical Modeling of Coal Washability, 1st Australia Coal Preparation Conf., April 6–10, Newcastle, Australia, 220–239.

Clarkson, C.J., 1992. Optimisation of Coal Production from Mine Face to Customer, 3rd Large Open Pit Mining Conference, Aug. 30–Sept. 3, Makcay, Australia, 433–440.

Dell, C.C., 1956. The Mayer Curve, *Colliery Guardian*, Vol. 33, pp. 412–414.

Mayer, F.W., 1950. A New Washing Curve. *Gluckauf*, 86: 498–509.

Peng, F.F. and Luckie, P.T., 1991. Process Control–Part I: Separation Evaluation, *Coal Preparation*, J. Leonard (Ed.), 5th Ed., SME, Littleton, Colorado, 659–716.

Rayner, J.G., 1987. Direct Determination of Washing Parameters to Maximize Yield at a Given Ash, Bull. Proc. Australia Inst. Mining and Metallurgy, 292(8): 67–70.

Rong, R.X. and Lyman, G.J., 1985. Computational Techniques for Coal Washery Optimization–Parallel Gravity and Flotation Separation, *Coal Preparation*, 2: 51–67.

Operating Characteristics of Water-only Cyclone/Spiral Circuits Cleaning Fine Coal

Peter Bethell[*] and Robert G. Moorhead[†]

Historically, either water-only cyclones or spirals have been used to clean 1.00 mm × 0.15 mm (16M × 100M) size coal in the U.S. Recently, there have been several commercial circuits installed which combine the water-only cyclone and spiral into a two-stage circuit. The different operating characteristics of each of these cleaning devices are ideally suited for this integration. This paper describes, through actual operating data, the flexibility of the water-only cyclone/ spiral circuit configuration in processing fine-size coal. One example shows how the circuit can be configured to provide maximum carbon recovery with low recycle rates. A second example shows how this circuit can permit operating the water-only cyclones at low D_{50}'s to produce a premium-quality product from a high-ash raw feed while minimizing coal loss through recleaning the water-only-cyclone underflow in second-stage spirals.

INTRODUCTION

Fine-coal, gravity-based cleaning has always required compromises because the typical performance associated with cleaning 1 mm × 0.25 mm (16M × 60M) size particles in water-based cleaning circuits is well below the performance achieved by dense-media processing. However, due to the difficulties in magnetite recovery, media control, and associated operating costs, dense media has not been successfully employed to clean fine coal.

Thus either water-only cyclones or spirals have typically been employed to clean the 1 mm × 0.25 mm (16M × 60M) size fraction. In order to improve their overall performance, it is common practice to configure them in two-stage circuits. The two basic types of two-stage cleaning circuits are:

---

[*] Massey Energy Co., Chapmanville, W.Va.

[†] Krebs Engineers, Tuscon, Ariz.

- Rewash circuits, which direct one of the first-stage streams to a second stage and the product from this stage is combined with the first-stage product.
- Recycle circuits, which direct one of the first-stage streams to a second stage and a stream from the second stage is recycled back to the first-stage feed.

ADVANTAGES AND DISADVANTAGES OF TWO-STAGE CIRCUITS

Each of these circuits has advantages and disadvantages. The rewash circuit requires less equipment than a recycle circuit because there is no stream recycling back to the raw feed to increase the overall tonnage requiring a larger primary-stage circuit. However the reclean circuit is not as efficient as a recycle circuit (1).

The advantage of the recycle circuit is that, because a second-stage stream is recycled, it provides improved performance over a rewash circuit. The disadvantage is the additional processing capacity required due to the recycling load.

The typical disadvantages of two-stage, water-only cyclone circuits is the requirement to install two pumps and sumps, as well as the added complexity in adjusting their separating densities. Whenever a two-stage, recycle circuit requires adjustment, the dynamics of recycling demand more analysis to determine which stage needs adjustment as well as the impact of that adjustment on the amount of solids being recycled.

Although a water-only cyclone/spiral circuit with recycle is no easier to adjust, the spirals can usually be gravity fed. This eliminates the need for a second pump and sump.

Historically, two-stage circuits have been comprised of identical equipment. This approach reduces spare-parts inventories because all of the equipment in the circuit requires essentially the same parts. The disadvantage is that the performance trait that created the separation error in the first stage is duplicated in the second stage.

A classic example of this is a conventional two-stage, middlings-recycle, water-only cyclone circuit, as shown in Figure 1. It is well known that water-only cyclones are "classifier" type equipment (2) with the undesirable characteristic of separating coarser-sized particles at lower D_{50}'s than finer-sized particles. If coarse-sized coal is fed to this circuit, the first-stage rejects the coarse-sized coal and directs it to the second stage. The second-stage, having the same separation trait, also rejects these particles. The result is a minimal benefit in a multiple-stage circuit.

ADVANTAGES OF THE WATER-ONLY CYCLONE/SPIRAL CIRCUIT

Recently, there has been a trend towards combining different types of equipment in a two-stage circuit—specifically, water-only cyclones and spirals. Figure 2 shows a generic water-only cyclone/spiral circuit. Although these two cleaning devices can be fed the same size range, their cleaning characteristics are very different.

As previously stated, water-only cyclones are "classifier" type devices and separate the 1mm × 28M size fraction at a lower D_{50} than the 0.595 mm × 0.25 mm (28M × 60M) size fraction. In contrast, a spiral is a "flowing-film" type separator and it, in contrast, separates the 1.0 mm × 0.595 mm (16M × 28M) size fraction at a higher D_{50} than the 0.595 mm × 0.25 mm (28M × 60M) size fraction. The result of combining these two different cleaning devices in a single circuit is that the spiral's cleaning characteristic compensates for the water-only cyclone's.

Besides the general symbiotic performance characteristics of the water-only cyclone and spiral, there are multiple other advantages as well:

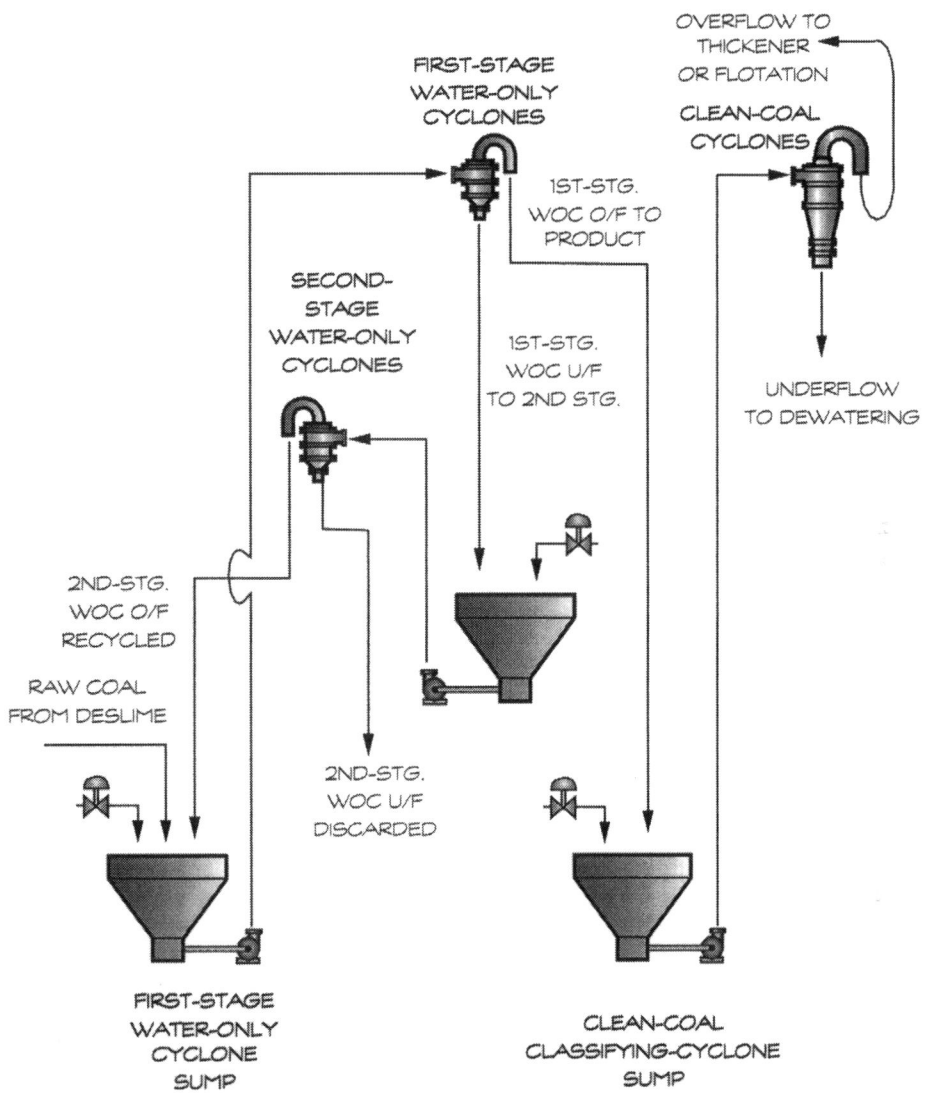

FIGURE 1 Typical two-stage, middlings-recycle water-only cyclone circuit

- Water-only cyclones can be configured to provide relatively low separating densities but usually at the expense of coal loss. The spiral, on the other hand, cannot provide as low a separating density as the water-only cyclone but, characteristically does not lose coal in the refuse split. Combining these two traits permit achieving a lower overall separating density without the coal loss of a typical two-stage, water-only cyclone circuit.
- Water-only cyclones cost roughly $325 per-ton capacity while spirals cost between $571 and $800 per-ton capacity. Using the lower capital-cost water-only cyclones as first-stage cleaners (which handles the highest tonnage) potentially reduces capital cost.

96 | COAL-BASED GRAVITY SEPARATIONS

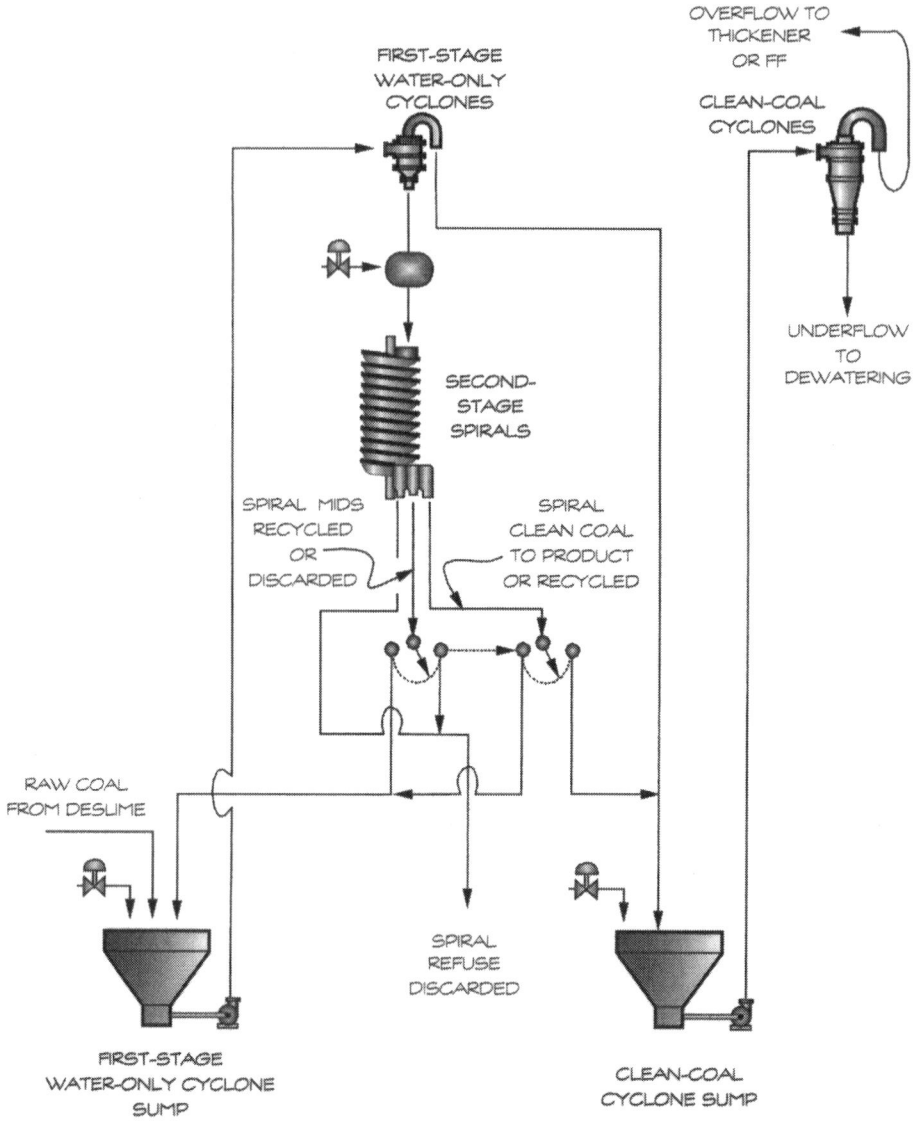

FIGURE 2 Typical water-only cyclone/spiral circuit with various recycle options

- In a typical two-stage, water-only cyclone circuit, the second-stage overflow is recycled back to the first-stage feed. Since the majority of the ~15 percent solids (w/w) feed slurry reports to the overflow of a cyclone, the first-stage capacity must be increased merely to handle the water, not the solids, being recycled from the second-stage, water-only cyclone overflow. Using spirals as the second-stage cleaning device provides much more concentrated recycle streams. Typically, the spiral clean-coal stream will discharge at between 18 to 25 percent solids (w/w) and the middlings will discharge at between 35 and 45 percent solids (w/w). The more concentrated spiral discharge streams require less first-stage hydraulic capacity to process.

- Due to the "classifier" type nature of the water-only cyclone, the underflow stream is well deslimed. This produces an optimum feed for the spiral because the water-only-cyclone underflow does not contain excessive slimes which can compromise the spirals performance.
- Spirals produce three products (clean coal, middlings, and refuse) in a relatively concentrated slurry. The existence of separate clean coal and middlings streams permits multiple recycle options:
  - No recycle
  - Middlings only recycle
  - Clean-coal only recycle
  - Clean-coal and middlings recycle

These multiple-recycling options permit utilizing the circuit regardless of whether the objective is to make mid- or high-density separation. Figure 2 also shows the various recycle options available and the potential to select a particular recycle mode while the circuit is online.

CASE NO. ONE: WATER-ONLY CYCLONE/SPIRAL WITH MIDDLINGS RECYCLE

Application

This water-only cyclone/spiral circuit was installed to specifically remove calcite in a 0.595 mm × 0 (28M × 0) feed. The objective was to make a high specific-gravity separation to ensure high carbon recovery while still achieving high calcite removals.

Figure 3 shows a schematic flow sheet of this water-only cyclone/spiral circuit. Because a high overall separating density was desired, the spiral clean coal was directed to product along with the water-only cyclone overflow and the middlings stream was recycled.

Performance

The overall performance of this water-only cyclone/spiral circuit achieved very acceptable performance, producing a plus × 0.15 mm (100M) size product with 7.03 percent ash, 1.18 percent sulfur at a yield of 91.43 percent. The recycle rate ranged from 12.6 percent for the plus 0.595 mm (28M) size to 8.10 percent for the 0.595 mm × 0.15 mm (28M × 100M) size and the overall recycle rate for total solids (plus × 0) was 10.2 percent.

The recycle rates for this water-only cyclone/spiral circuit were less than half that of a typical two-stage, middlings-recycle, water-only cyclone circuit. The reason for this significantly lower recycle rate is related to only the spiral middlings being recycled. The three-product discharge of a spiral permits directing the spiral clean coal to product, while a second-stage water-only cyclone (with only a two-product discharge) gives no option but to recycle the entire overflow which contains the clean-coal and middlings fractions. The presence of the clean-coal fraction in the water-only cyclone overflow results in the significantly higher recycle rates in comparison to a second-stage spiral.

The plus 0.15 mm (100M) size fraction organic efficiencies were above 99 percent while recovering 97.8 percent of the combustibles and removing 50.94 percent of the ash (also shown in Table 1).

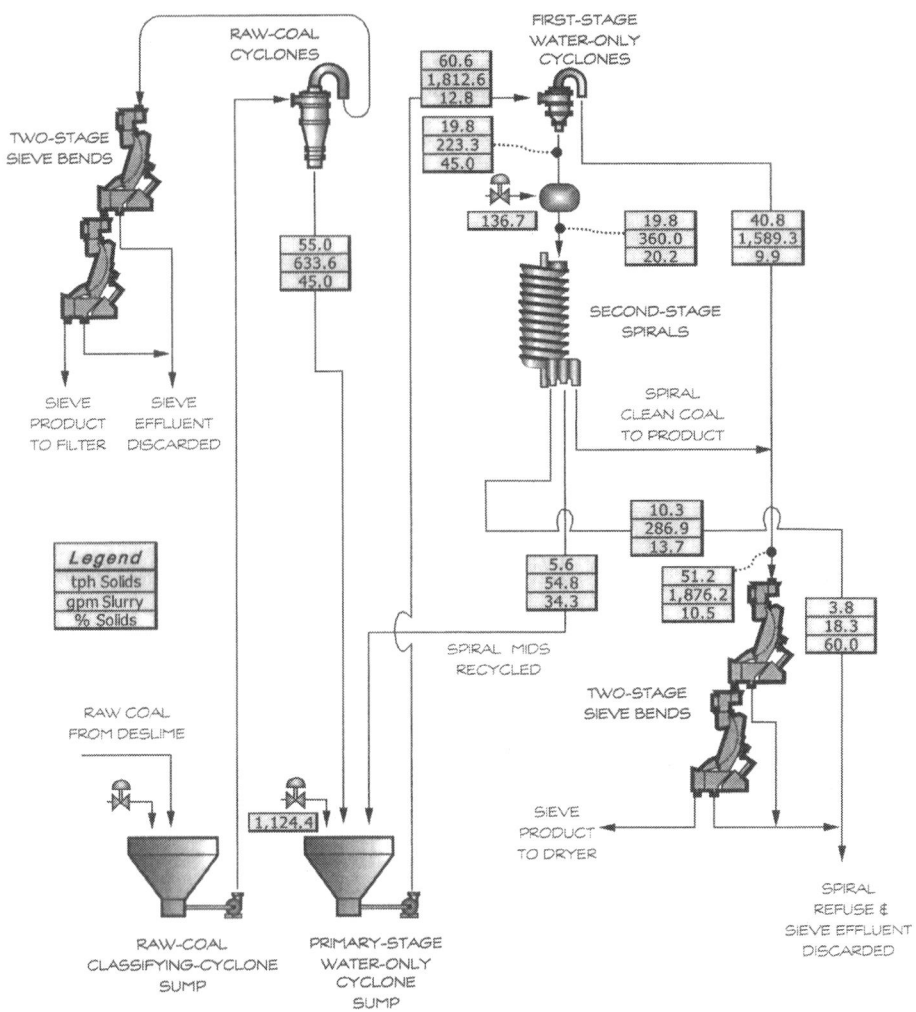

FIGURE 3 Case No. 1: water-only cyclone/spiral circuit with middlings recycled

Figure 4 shows the recovery curves for the plus 0.595 mm (28M), 0.595 mm × 0.15 mm (28M × 100M), and plus 0.595 mm × 0.15 mm (28M × 100M) composite size fraction. What is most indicative of the advantages of combining water-only cyclones with spirals is the overall D_{50}'s of the plus 0.595 mm (28M) and 0.595 mm × 0.15 mm (28M × 100M) size fractions. The plus × 0.595 mm (28M) size D_{50} is 2.04 SG while the 0.595 mm × 0.15 mm (28M × 100M) size D_{50} is 2.12 SG—or a differential of only 0.08 SG. This low differential in D_{50}'s, between size fractions, is one reason for the exceptional performance of this circuit configuration.

It must be remembered that optimum gravity-separation performance is predicated on incremental ash recovery. The closer the D_{50}'s are between sizes, the closer their respective theoretical recovered incremental ash values will be. Likely, nothing short of dense-media processing could theoretically achieve this close a D_{50} between size fractions.

TABLE 1 Performance values. Case No. 1: Water-only cyclone/spiral circuit with middlings recycled

Size Fraction	Plus 28M	28M × 100M	Plus × 100M
Product			
Yield (wt%)			
WOC	48.20	76.57	69.55
Spiral	44.02	14.64	21.67
Overall	92.21	91.21	91.43
Ash (wt%)			
WOC	4.80	6.35	6.01
Spiral	10.49	10.08	10.28
Overall	7.41	6.95	7.03
Sulfur (wt%)			
WOC	1.00	1.08	1.08
Spiral	1.37	1.76	1.59
Overall	1.17	1.19	1.18
Refuse			
Yield (wt%)	7.79	8.79	8.57
Ash (wt%)	80.69	76.93	77.86
Sulfur (wt%)	6.72	9.57	8.87
Separating Density			
WOC	1.33	1.61	1.52
Spiral CC	1.63	1.52	1.56
Spiral Refuse	2.31	2.27	2.29
Overall	2.04	2.12	2.09
Probable Error			
WOC	0.121	0.260	0.198
Spiral CC	0.162	0.121	0.136
Spiral Refuse	0.166	0.199	0.172
Overall	0.156	0.254	0.210
Overall Circuit			
Recycle Rate (wt%)	112.60	108.10	108.91
Organic Efficiency	99.75	99.63	99.64
Ash Removal (wt%)	47.90	51.61	50.94
Sulfur Removal (wt%)	32.61	43.64	40.90
Combustibles Recovery (wt%)	98.27	97.67	97.82

(1) Overall recycle rate for plus × 0 = 110.2.

In contrast, the D_{50}'s for the water-only cyclone were 1.33 SG for the plus × 0.595 mm (28M) and 1.61 SG for the 0.595 mm × 0.15 mm (28M × 100M) size fraction—a 0.28 SG differential (or 3.5 times greater than the overall circuit D_{50}'s for the same size fractions). The fact that the spiral separates the plus 0.595 mm (28M) at a higher D_{50} than the 0.595 mm × 0.15 mm (28M × 100M) compensates for the water-only cyclone's inversed response.

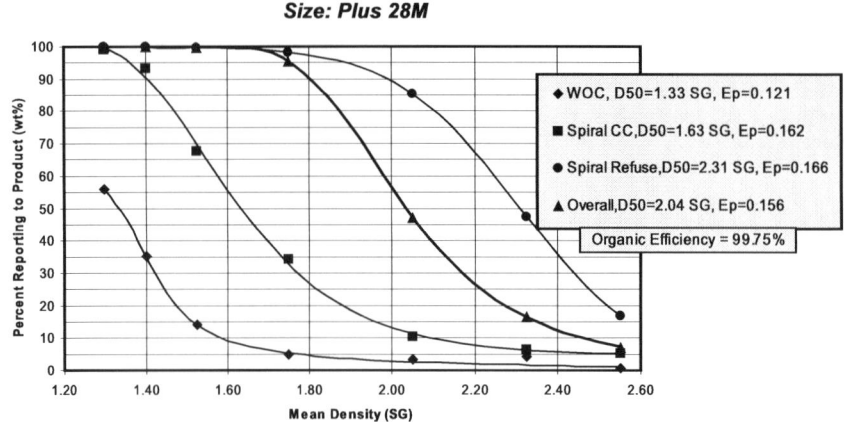
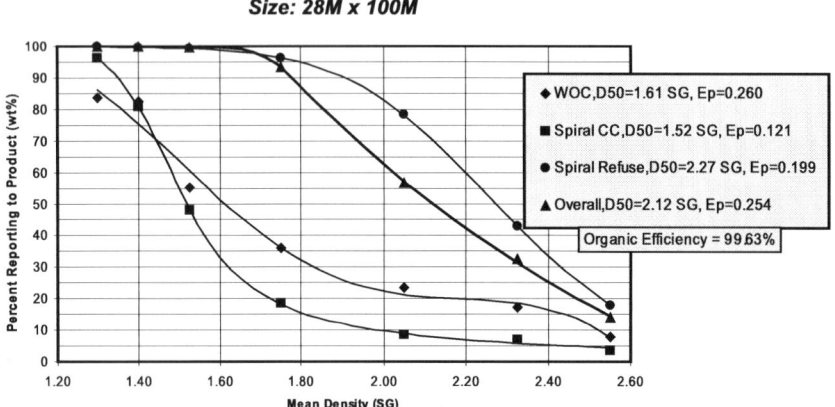
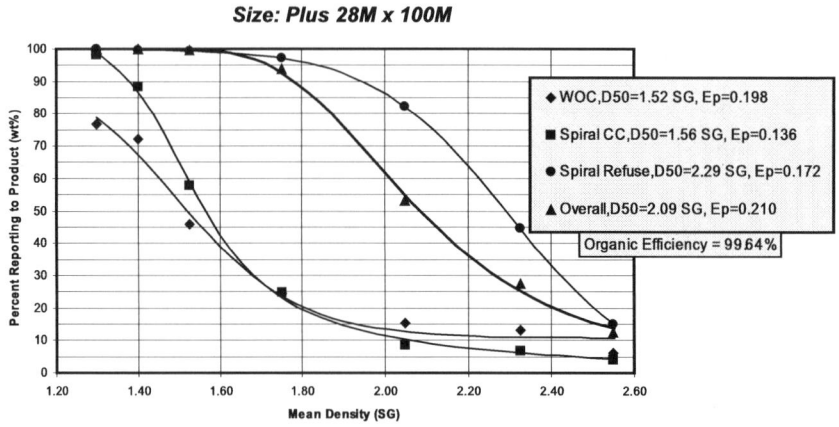

FIGURE 4 Recovery curves for Case No. 1: Water-only cyclone/spiral circuit with middlings recycled

CASE STUDY NO. TWO: WATER-ONLY CYCLONE/SPIRAL WITH NO RECYCLE

Application

This particular water-only cyclone/spiral circuit is presently operating at Massey Energy's Bandmill Plant in West Virginia. It is employed to clean a very high-ash raw coal (37.1 percent ash in the 1mm × 0.15 mm (16M × 100M) size). The water-only cyclone feed also contains a high percentage of clay (44.31 wt % 0.15 mm × 0 (100M × 0) at 58.21 percent ash). The circuit configuration, as shown in Figure 5, differs from Case No. One in that no spiral stream is recycled back to the water-only-cyclone feed. The high-ash of the feed potentially limits recycle in this case because the overall separation density would increase unacceptably and result in too high a product ash.

Performance

Table 2 lists pertinent performance values for the various streams. The implementation of spirals as second-stage cleaners permits operating the water-only cyclones at very low D_{50}'s, without concern about coal losses in the underflow. As shown in Table 2, the water-only cyclone produced a plus × 0.15 mm (100M) size product with 5.17 percent ash at a 49.62 percent yield. Given the high feed ash of the plus 0.15 mm (100M) size fraction (37.10 percent), the water-only cyclone provided a significant level of upgrade as well as acceptable yield on such a high-ash feed.

Due to the intentionally low operating D_{50} of the water-only cyclone to produce a very low product ash, the underflow contains too high a percentage of 1.60 SG float material to discard to refuse. The use of spirals as second-stage cleaners permits the recovery of a high percentage of this desirable coal in the water-only cyclone underflow, while producing a very high-ash refuse. Although the spirals have a yield of only 19.83 percent on a unit basis, they are recovering an additional 12.8 tph of 7.77 percent ash coal. This results in a composite plus × 0.15 mm (100M) size circuit product of 5.59 percent ash at an overall yield of 59.63 percent.

The combined middlings and refuse streams of the spirals, which represent the composite circuit, has an ash on the plus × 0.15 mm (100M) size of 83.67 percent.

It is unlikely that any other commercially available fine-coal-cleaning circuit would be capable of recovering 65.52 percent of the plus 0.15 mm (100M) size at such a low-ash while at the same time producing a refuse with over 83 percent ash.

It should be pointed out that the performance initially achieved by this circuit, after commissioning, did not meet Massey Energy's expectations. The present performance of this circuit required a cooperative effort between the equipment supplier and Massey Energy to determine acceptable operating conditions and orifice combinations to provide the product quality and yield presently being achieved by this circuit.

In summary, the advantages of the water-only cyclone/spiral circuit in this case are:

- The implementation of spirals as the second-stage cleaners, with their separation characteristic of always producing a reject with little carbon loss, permits operating the water-only cyclones at very low D_{50}'s. This provides a very low-ash product from a very high-ash feed.
- Water-only cyclones provide a lower D_{50} than could be achieved by any other water-based cleaning device.
- The water-only cyclones are effective "deslimers" and remove significant amounts of clay from the spiral feed. While the water-only-cyclone feed contains

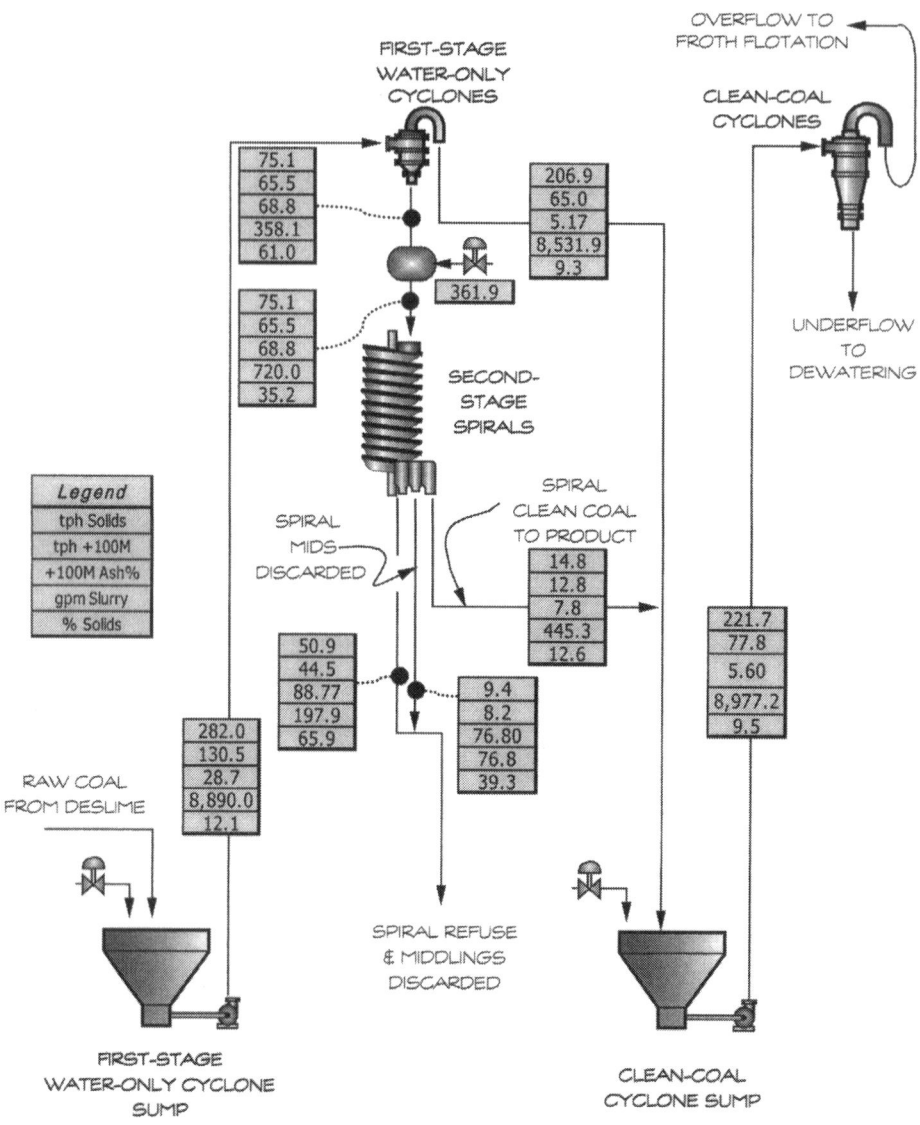

FIGURE 5 Case No. 2, water-only cyclone/spiral circuit with no recycle

44.31 percent minus 0.15 mm (100M) size, the spiral feed contains only 12.28 percent minus 100M size.

PRACTICAL APPLICATION OF WATER-ONLY CYCLONE/SPIRAL CIRCUITS

Although the combination of water-only cyclones and spirals has many benefits, there are some guidelines that need to be followed when applying it to fine-coal cleaning. Because the water-only cyclone and spiral separation densities vary so greatly with size, there is a potential for unacceptably high recycle rates if an inappropriate recycle configuration is selected.

TABLE 2 Performance values. Case No. 2: Water-only cyclone/spiral circuit with no recycle

Size Fraction	Plus 28M	28M × 100M	Plus 1.60 mm	1.60 mm × 0.75 mm	0.75 mm × 100M	Plus 100M
Feed						
Weight (wt%)						
WOC	17.17	38.52				55.69
Spiral			1.41	21.28	65.03	87.73
Ash (wt%)						
WOC	33.77	38.59				37.10
Spiral			63.39	53.89	73.56	68.62
Product						
Yield (wt%)						
WOC	47.01	52.35				49.82
Spiral-unit basis			30.81	38.14	13.28	19.83
Spiral-circuit basis						9.81
Overall						59.63
Weight (wt%)						
WOC	4.59	26.82				31.41
Spiral			2.21	42.34	42.77	87.32
Overall						
Ash (wt%)						
WOC	3.22	5.50				5.17
Spiral			13.35	7.20	8.04	7.77
Overall						5.59
Refuse						
Yield (wt%)						
Spiral-unit basis			69.19	60.86	86.72	80.17
Spiral-circuit basis						40.38
Overall						40.38
Weight (wt%)						
Spiral			1.22	16.12	70.48	87.82
Ash (wt%)						
Spiral			85.68	83.92	83.59	83.68

As a general rule, the 1.0 mm × 0.15 mm (16M × 100M) size fraction can be successfully cleaned in a 15-inch Ø water-only cyclone with no recycle or recycling spiral-middlings only. Recycling the spiral clean coal or combined clean-coal and middlings can potentially create unacceptably high recycle rates with a nominal feed top-size of 1 mm (16M).

When cleaning a nominal 0.595 mm (28M) top size in a 15-inch Ø water-only cyclone/spiral circuit, normally the two options of recycling spiral-middlings only or spiral-clean-coal only can be selected without concern about excessive recycle rates. Recycling combined spiral clean-coal and middlings requires maintaining sufficiently

TABLE 3 Float/sink data by stream. Case No. 2: Water-only cyclone/spiral circuit with no recycle

First-stage Water-only Cyclone. Size: Plus 28M × 100M

	Float 1.60 SG	Sink 1.60 SG	Head
Feed			
Weight (wt%)	60.82	39.18	100.00
Ash (wt%)	4.56	86.20	36.55
Overflow			
Weight (wt%)	97.63	2.37	100.00
Ash (wt%)	3.62	73.58	5.28
Underflow			
Weight (wt%)	23.59	76.41	100.00
Ash (wt%)	7.43	86.51	67.86

Second-stage Spirals. Size: Plus 1.60 mm × 100M

	Float 1.60 SG	1.60 × 1.80 SG	Sink 1.80 SG	Head
Clean Coal				
Weight (wt%)	95.05	2.64	2.21	100.00
Ash (wt%)	5.75	41.64	66.74	8.05
Middlings				
Weight (wt%)	34.63	10.80	54.57	100.00
Ash (wt%)	12.27	42.88	77.35	51.09
Refuse				
Weight (wt%)	1.31	0.75	97.94	100.00
Ash (wt%)	13.15	47.88	89.86	88.54

high water-only-cyclone D_{50}'s to avoid excessive recycle rates. That "minimum" water-only-cyclone D_{50} will be determined by the specific washability characteristics of the raw feed being processed.

In order to demonstrate the potential recycle rates of a water-only cyclone/spiral circuit, the cleaning of a hypothetical 1.0 mm × 0.15 mm (16M × 100M) size raw-coal feed was modeled in a water-only cyclone/spiral computer simulation program. The three (3) recycle options of:

- Spiral middlings only,
- Spiral clean-coal and middlings, and
- Spiral clean coal

were simulated while varying the water-only cyclone overall D_{50} from 1.40 to 1.70 SG, and the calculated recycle rates were recorded.

Figure 6 shows the recycle-rate characteristics for the 1.0 mm × 0.15 mm (16M × 100M), 1.0 mm × 0.595 mm (16M × 28M), and 0.595 mm × 0.15 mm (28M × 100M) size fractions. The 1.0 mm × 0.595 mm (16M × 28M) size recycle rate will always be the determining variable because of the greater D_{50} differentials between the water-only cyclone and spiral for this size fraction.

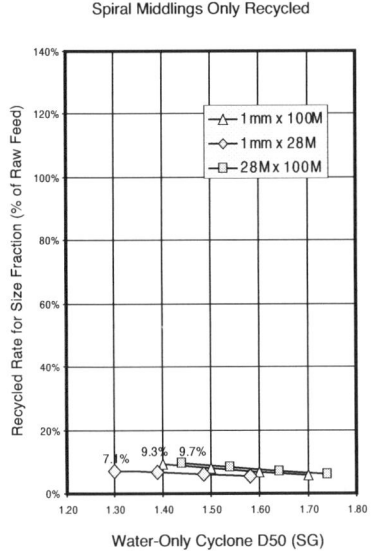
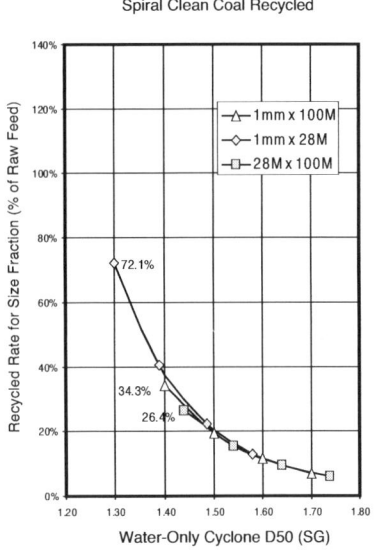
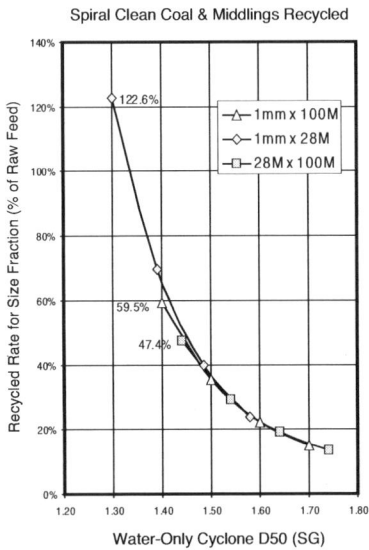

FIGURE 6 Response of recycle rates to water-only cyclone D_{50}'s for different recycle configurations

As shown in the upper-left chart in Figure 6, when recycling only spiral middlings, the recycle rate never rose above 7.1 percent of the raw-feed rate for the 1.0 mm × 0.595 mm (16M × 28M) size fraction when simulating the water-only cyclone at its lowest overall D_{50} (1.40 SG). The graph shows the recycle rate for all size fractions to be nearly constant, regardless of the water-only cyclone's separation density. It also illustrates the advantage of a water-only cyclone/spiral circuit, operated with middlings recycle, over a two-stage, water-only cyclone circuit because a typical two-stage, middlings-recycle, water-only cyclone circuit would have recycle rates of between 20 and 25 percent of the raw-feed rate.

However, when recycling spiral clean coal, the water-only cyclone/spiral circuit shows a propensity to substantially increase the recycle rate of the 1.0 mm × 0.595 mm (16M × 28M) size fraction, as shown in the upper-right chart in Figure 6. The 1.0 mm × 0.595 mm (16M × 28M) size recycle rates rises above 40 percent of the raw-feed rate when the water-only cyclone's overall D_{50} decreases below 1.50 SG and rises to 72.1 percent of the raw-feed rate when the water-only cyclone D_{50} is decreased to 1.40 SG. These represent unacceptably high recycle rates that potentially compromise the unit operation of the cleaning devices.

However, the 0.595 mm × 0.15 mm (28M × 100M) size recycle rates, in this mode, never rise above 26.4 percent of the raw-feed rate, even when the overall D_{50} of the water-only cyclone is decreased to 1.40 SG. This recycle rate, for the 0.595 mm × 0.15 mm (28M × 100M) size fraction, is roughly equivalent to typical recycle rates of middlings-recycle, water-only cyclone circuit. For this reason, recycling spiral clean-coal should only be selected if the nominal feed top-size is 0.595 mm (28M).

The lower chart in Figure 6 demonstrates the recycle rates when recycling both the spiral clean-coal and middlings. This chart indicates that this recycle option should only be selected if the water-only cyclone D_{50} is maintained above approximately 1.55 SG. Water-only cyclone D_{50}'s below this density will result in prohibitively high recycle rates—approaching 122.6 percent of the raw feed for the 1.0 mm × 0.595 mm (16M × 28M) size fraction.

This hypothetical study indicates why it is so important to take into account the water-only cyclone's anticipated separating density as well as the planned circuit nominal top-size in selecting the recycle mode for the water-only cyclone/spiral circuit.

CONCLUSIONS

The water-only cyclone/spiral circuit can be used to clean fine-size coal with exceptionally high efficiency. The presented operating data demonstrates how flexible and configurable this circuit can be to operate at a high separating density, to maximize carbon recovery, or alternately be configured to operate at a low separating density, to provide a premium-grade product.

The implementation of the circuit, however, requires some diligence to ensure excessive recycle rates do not result. Essentially, when feeding a nominal 1.0 mm (28M) top-size, the circuit must only be configured with no recycle or spiral middlings only recycled. When cleaning a nominal 0.595 mm (28M) top-size, spiral middlings and clean coal can typically be recycled without concern of excessive recycled rates. Recycling both the spiral clean-coal and middlings must be restricted to those nominal 0.595 mm (28M) top-size-feed applications where the water-only cyclone separating density is sufficiently high enough to keep the recycle rate acceptable.

REFERENCES

Journal Articles

1. Laverick, M.K., 1969, "Theoretical Evaluation of Two-Stage Washing and Middlings-Recirculation Systems," *Coal Preparation,* March/April, pp. 55–59.
2. Weyher, L.H.E., Lovell, H.L., 1969, "Hydrocyclone Washing of Fine Coal," *Transactions– Vol. 244, Society of Mining Engineers, AIME,* June, pp. 191–203.

Comparing a Two-stage Spiral to Two Stages of Spirals for Fine Coal Preparation

Peter J. Bethell[*] and Barbara J. Arnold[†]

Many coal cleaning plants use two stages of coal spirals to upgrade the 1 mm × 0.15 mm size fraction. Middlings recirculation is sometimes included in this circuitry. Spirals that combine both stages into one unit—a rougher/cleaner with the capability of middlings recycle—are now available. Following in-plant testing at an existing plant that uses two stages of spirals, plant operators chose the new spiral for their plant addition. This paper provides a comparison of the performance of the existing two-stage spiral circuit with this new spiral.

INTRODUCTION

Independence Coal Co., located in Madison, WV, began plans for an 800 tph stand-alone plant addition in 2001 for their Liberty Processing Plant. As part of this planning, a two-stage SX7 test spiral manufactured by Multotec Process Equipment, Kempton Park, South Africa, was tested and compared to the two-stage middling retreatment spiral circuit in the old plant. Following this testwork, the SX7 spiral was selected for inclusion in the new plant addition circuitry based on better performance with the same feed material. Following commissioning in 2002, operating plant data comparing the old and new spiral circuits showed that the SX7 spiral circuit was performing considerably better than the existing two-stage spiral circuit.

BENEFITS OF DOUBLE STAGES OF SPIRALS

Coal spiral circuits consist of either a single stage or a double stage of spirals. Circuits incorporating primary middlings retreatment have been shown to be more efficient than

---

[*] Massey Coal Services, Inc., Chapmanville, W.Va.

[†] PrepTech, Inc., Apollo, Pa.

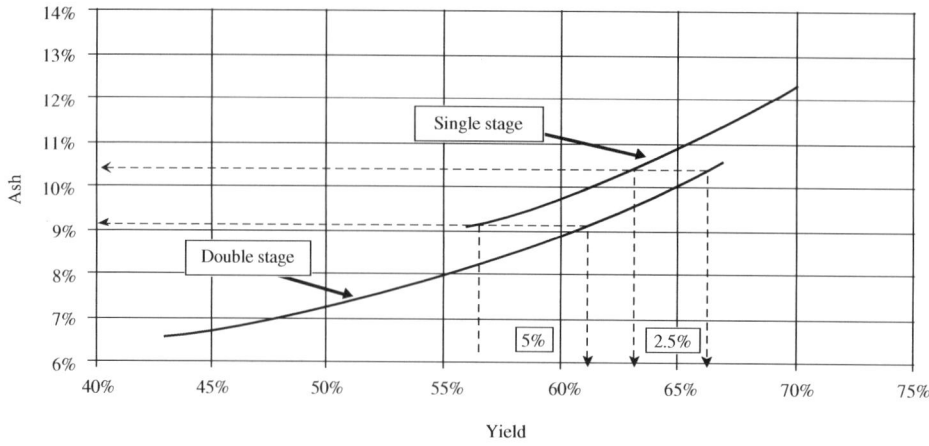

FIGURE 1 Performance of a single stage spiral circuit versus a double stage spiral circuit for an easy to clean coal

single stage circuits (Bethell, Stanley, and Horton, 1991; Bethell, 2002). A double stage configuration could be used for retreatment of a primary middlings or of the total primary clean coal/middlings product—which is the case for more difficult-to-wash coals. Figure 1 shows a comparison of a single stage and double stage spiral circuit for the 14 × 60 mesh (1.2 × 0.25 mm) size fraction of an easy-to-wash coal (less than 2 percent near-gravity material). For the same clean coal ash content, yield improvements of between 2.5 and 5 percent can be achieved (Prinsloo and Abela, 1998).

However, a double stage circuit is more costly than a single stage circuit in terms of capital expenditures and operating costs. The development of a single spiral with the capabilities of a double stage spiral circuit was initiated to reduce this cost while maintaining or improving performance.

DESCRIPTION OF THE SX7 TWO-STAGE SPIRAL

The result of this development work was the MX7 and SX7 spiral. Four spiral turns are followed by the removal of a primary refuse and re-mixing of the middlings and clean coal. No water is added to the remixer (or repulper) as sufficient water flows with the clean coal and middlings from the first four turns. The middlings and clean coal are then treated on three additional spiral turns. A conventional splitter at the base of the spiral allows removal of a secondary refuse that can be combined with the primary refuse. A middlings can also be collected and fed back to the feed for re-processing or can be combined with the clean coal product.

An MX7 spiral was first developed for use with difficult-to-clean coals with a high percentage of near-gravity material. The SX7 spiral was developed for use with relatively easy-to-clean coals, such as the coal treated at Liberty. Attention was paid to the reject capacity of the SX7 primary refuse splitter. As shown in Figure 2, a coal sample that was spiked with pure sand was tested. The refuse splitter was still capable of handling this material.

The benefits of the two-stage spiral compared to double stages of spirals are in reduction of total capital equipment costs, with a reduction in overall plant height and floor area as illustrated in Figure 3. Reduction in operating costs are also achieved as

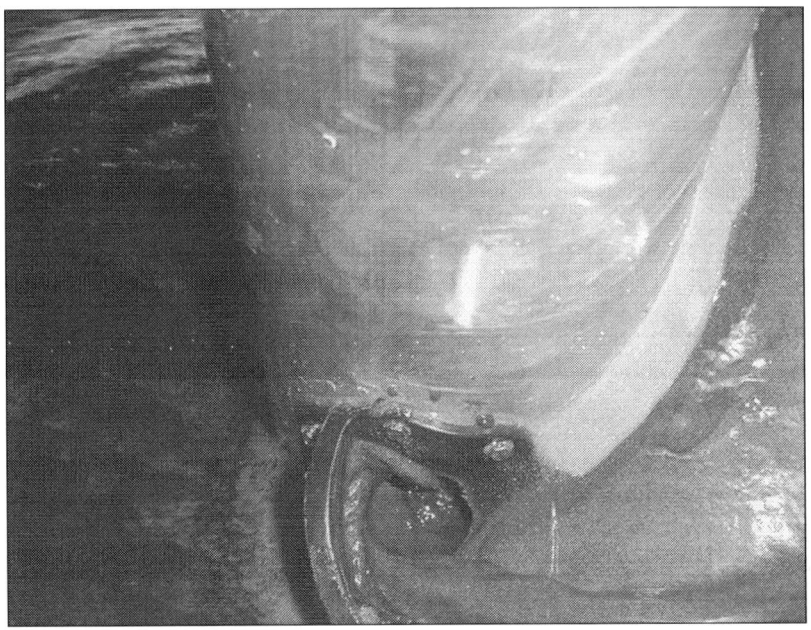

FIGURE 2 SX7 refuse splitter rejecting sand

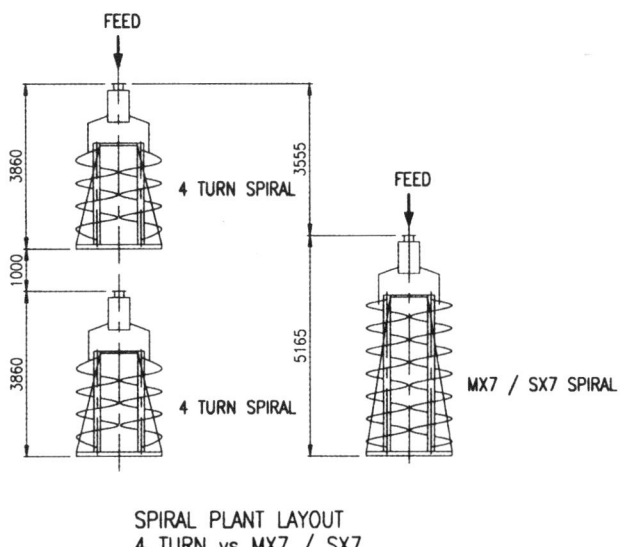

FIGURE 3 Layout of a two-stage spiral circuit compared to one stage of an SX7 circuit

only one set of spirals need be adjusted, rather than two. There are also advantages in reducing errors that are caused during slurry distribution by eliminating one stage of slurry distribution.

INITIAL TEST WORK—SX7

In August 2001, a single turn SX7 test spiral was tested at the Liberty Processing Plant using a slip stream of existing spiral feed. One test was conducted at 30 percent solids (by weight) in the feed and a second test at 25 percent solids (by weight). Yield versus ash content is given for each test by size fraction in Figures 4 and 5. Figure 4, at 30 percent feed solids, shows yield of around 85 percent for each of the plus 100 M (0.15 mm) size fractions with clean coal ash contents of about 5 percent. Figure 5, at 25 percent feed solids, shows yields of about 70 percent at similar ash levels. The total yield versus ash curves in each figure include the minus 100 M (0.15 mm) fraction, which constitutes approximately 27 percent of the feed to the spirals for these tests.

The performance of the spiral for the 16 × 100 M (1 × 0.15 mm) fraction is of greatest interest, as the clean coal is cycloned further to remove additional minus 100 mesh (0.15 mm). Partition curves for the clean coal and middlings for each test are given in Figures 6 and 7. The probable errors given are not adjusted for separating gravity. Typical normalized probable errors for one stage of a individual four-turn spiral trough for this particle size range are generally about 0.20.

OPERATING PLANT COMPARISON

The new plant was commissioned in April and May 2002, with 12 triple start SX7 spirals installed with a designed feed capacity of 2.5 stph/start. Middlings are recycled to the feed in this plant. The two-stage spiral circuit in the old plant consists of two stages of four-turn large diameter coal spirals with only the middlings from the first stage retreated in the second stage of spirals. While reams of data are available for comparison of the old and new spiral circuits, a representative data set for each circuit is used for a detailed comparison here as given in Table 1. These samples were not collected on the same day, but are given to also indicate the variation in feed quality to the plants.

A comparison of the feeds to each spiral circuit is warranted as they are quite different. The ash content of the feed to the old plant is 39 percent ash, while the ash content of the feed to the new plant is 45 percent. The feed to the new spiral circuit contains more plus 100 mesh (0.15 mm) material and this material is of poorer quality. A 41 percent ash in the plus 100 M (0.15 mm) for the new plant is compared to a 31 percent ash in this fraction for the old plant; and the fraction that floats at a 1.6 s.g. is only 55 percent in the new plant feed as compared to 68 percent in the old plant feed. As expected, with poorer feed quality, the yield for the new spiral circuit is lower at 57 percent, as calculated using an ash balance technique, while the yield is 69 percent in the old spiral circuit. Clean coal ash levels for the plus 100 M (0.15 mm) fraction are about 7 percent—7.11 in the old plant versus 6.83 percent in the new plant.

To take the effects of the differences in the feed quality out of the comparison, the distribution to clean coal was calculated for each of the gravity fractions. In the new plant spiral circuit, the 1.6 float fraction recovery is essentially 100 percent, while it is about 98 percent in the old plant. Further, the recovery of tailings material (1.8 sink) is reduced in the new plant to slightly more than one percent from almost 6 percent in the old plant circuit.

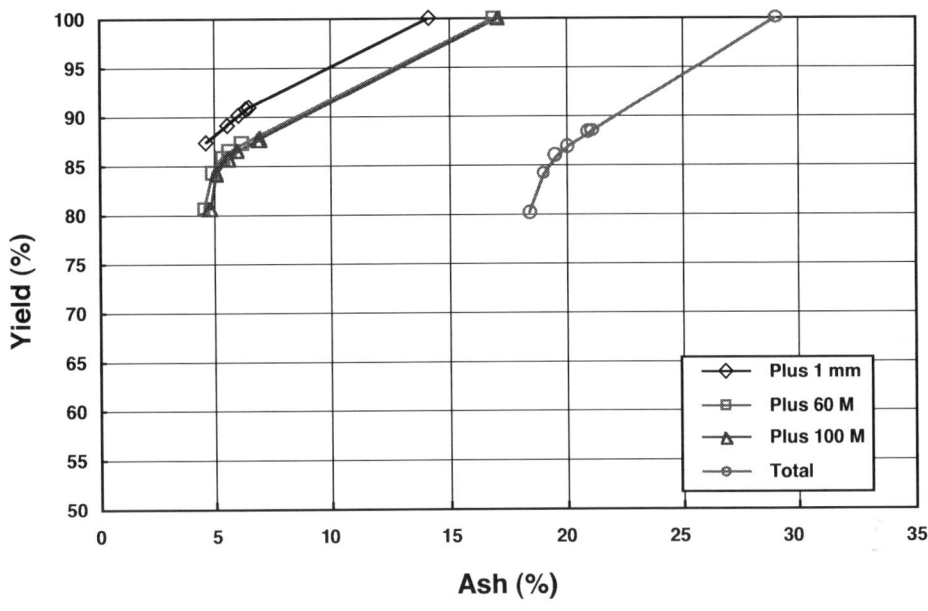

FIGURE 4 Yield versus ash content for various size fractions—SX7 Test No. 1

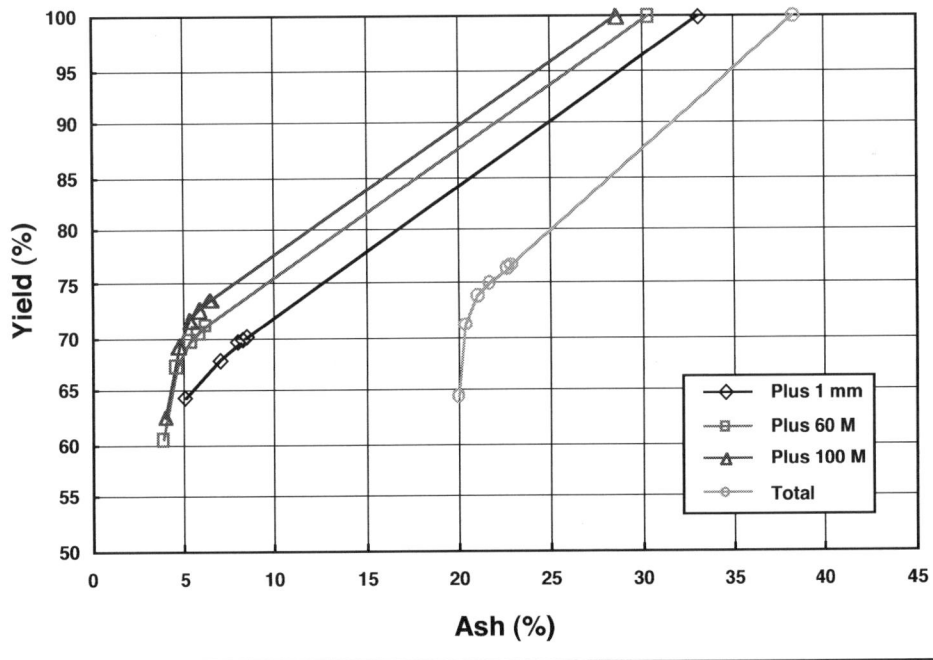

FIGURE 5 Yield versus ash content for various size fractions—SX7 Test No. 2

112 | COAL-BASED GRAVITY SEPARATIONS

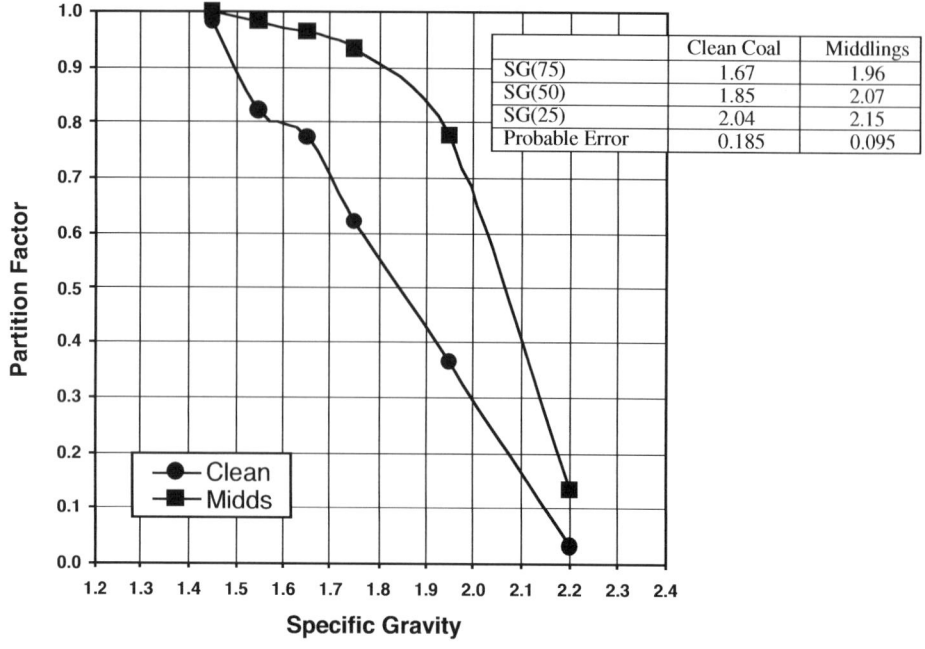

FIGURE 6   Partition curve for SX7 Test No. 1 (1 × 0.15 mm) at 30 percent (by weight) feed solids

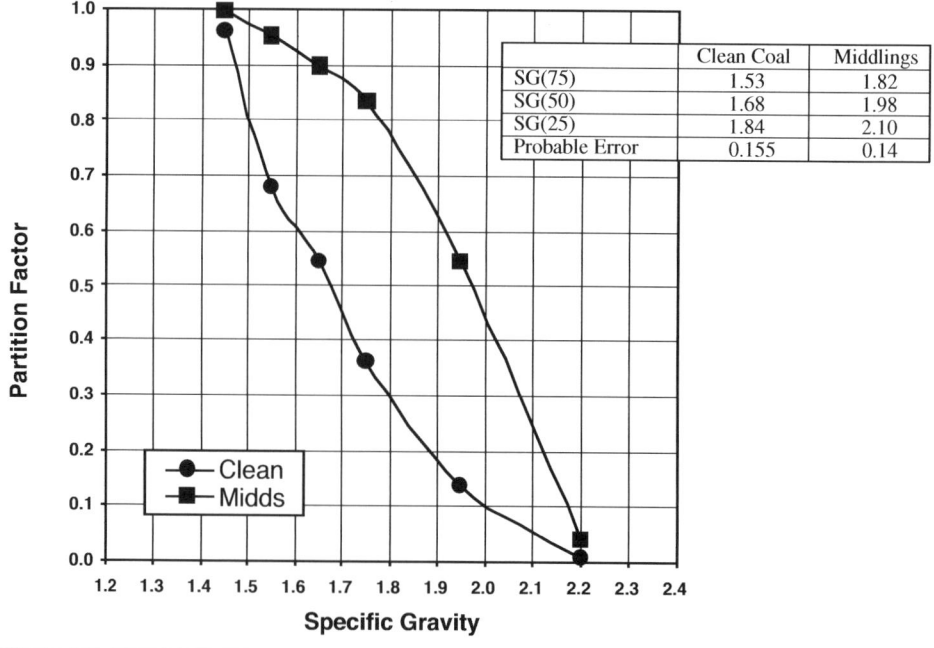

FIGURE 7   Partition curve for SX7 Test No. 2 (1 × 0.15 mm) at 25 percent (by weight) feed solids

TABLE 1 Performance comparison: Old plant spiral circuit with the new plant spiral circuit

Old Plant Spiral Circuit—Two-stage with middlings retreated in second stage

Size	Spiral Feed		Spiral Clean Coal		Spiral Refuse		Yield to Clean Coal (Wt %)
	Wt %	Ash (Wt %)	Wt %	Ash (Wt %)	Wt %	Ash (Wt %)	
+100 M	78.36	31.16	91.18	7.11	96.29	85.30	69.24
−100 M	21.64	65.96	8.82	54.46	3.71	82.14	58.45
Composite	100.00	38.69	100.00	11.28	100.00	85.18	62.91
Plus 100 M							Distribution to Clean Coal (%)
1.6 Float	68.21	6.81	96.53	4.87	3.87	9.35	97.99
1.6 × 1.8	2.75	42.52	1.01	40.15	2.80	44.41	25.43
1.8 Sink	29.05	88.29	2.46	70.61	93.33	89.22	5.86
Composite	100.00	31.46	100.00	6.84	100.00	84.88	

New Plant Spiral Circuit—SX7 spirals with overall middlings recycled to feed

Size	Spiral Feed		Spiral Clean Coal		Spiral Refuse		Yield to Clean Coal (Wt %)
	Wt %	Ash (Wt %)	Wt %	Ash (Wt %)	Wt %	Ash (Wt %)	
+100 M	85.88	40.92	96.50	6.83	92.01	86.91	56.72
−100 M	14.12	70.60	3.50	42.58	7.99	83.46	31.47
Composite	100.00	45.11	100.00	7.12	100.00	86.64	52.22
1 mm × 100M							Distribution to Clean Coal (%)
1.6 Float	55.17	5.95	97.30	4.69	1.40	13.86	100.00
1.6 × 1.8	1.72	36.40	1.65	33.90	0.96	43.11	54.41
1.8 Sink	43.11	86.00	1.03	57.23	97.63	88.26	1.36
Composite	100.00	40.98	100.00	5.71	100.00	86.77	

SUMMARY AND CONCLUSIONS

A spiral has been developed that incorporates two stages of spiraling into one assembly. For new plant installations, this new spiral has the advantages of reduced plant height and floor space requirements. Capital and operating costs are also reduced compared to conventional spiral circuitry that uses two stages of spirals. Performance of this new spiral (incorporating middlings recycle) is better than two stages of conventional spirals.

ACKNOWLEDGMENTS

The authors acknowledge the assistance of Dr. Gerald Luttrell, Virginia Polytechnic Institute and State University, for his assistance in data reduction.

REFERENCES

Bethell, P., Stanley, F., and Horton, S. 1991. Benefits Associated with Two-stage Spiral Cleaning at McClure Preparation Plant. *SME Annual Meeting*, Denver.

Bethell, P. 2002. Fine Coal Cleaning at Massey Energy. *19th Annual International Coal Preparation Exhibition and Conference*, Lexington.

Prinsloo, T.R., and Abela, R.L. 1998. Multiple Stage Fine Coal Spiral Concentrators. *Proceedings of the International Coal Preparation Congress*, Brisbane.

Advances in Teeter-bed Technology for Coal Cleaning Applications

Jaisen N. Kohmuench[*], Michael J. Mankosa[*], and Rick Q. Honaker[†]

Hindered-bed separators have been utilized for decades without many significant advances in the fundamental technology. Recently, there has been renewed interest in both the United States and Australia in using teeter-bed density separators as coal cleaning devices for the size range typically treated in spiral circuits. In response to this increased interest, Eriez developed the CrossFlow teeter-bed separator that overcomes some of the limitations associated with older designs while matching present-day production requirements. Data from laboratory- and pilot-scale evaluations, modeling investigations, and full-scale installations indicate that this new teeter-bed separator offers an improved capacity, a high organic efficiency, and many operational advantages.

INTRODUCTION

Hindered-bed hydraulic separators have been used in mineral processing applications for years. Simply stated, a hindered-bed separator is a vessel in which feed settles against an evenly distributed upward flow of water or other fluidizing medium. Typically, these devices are used for size classification; however, if the feed size distribution is within acceptable limits, hindered-bed separators can be used for the concentration of particles based on differences in density.

A simplified schematic of a typical hindered-bed separator is shown in Figure 1. Most hindered-bed separators utilize a downcomer to introduce feed material to the system. This material enters the feed zone and may encounter either free or hindered-settling conditions, depending on the concentration of particles in the separator. The settling particles form a fluidized bed (teeter bed) above the fluidization water injection point. Material is then segregated based on terminal, hindered-settling velocities. Slower

---

[*] Eriez, Erie, Pa.

[†] Dept. of Mining Engineering, Unviersity of Kentucky, Lexington, Ky.

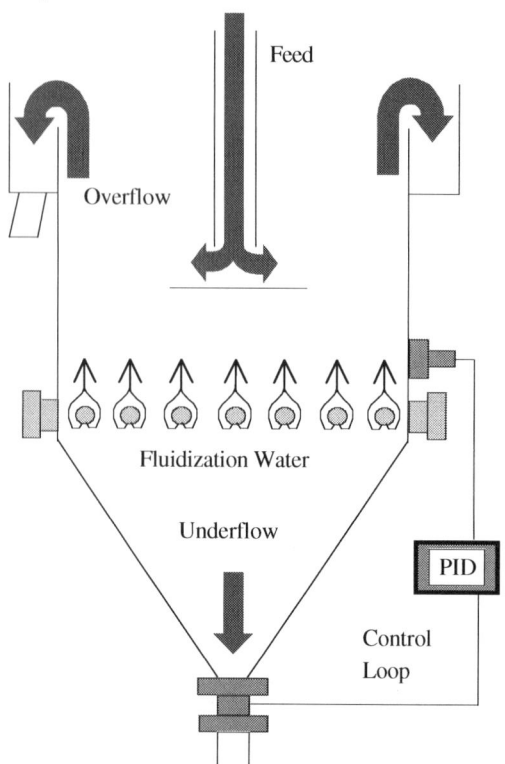

FIGURE 1 Schematic of a conventional hindered-bed separator

settling material reports to the top of the teeter bed while the faster settling particles descend to the bottom of the teeter zone. Specifically, low-density and fine material report to the overflow, while coarse and high-density material report to the underflow. Particles that settle through the teeter bed are discharged through an underflow control valve. The rate of underflow discharge is generally regulated using a PID control loop.

INNOVATIONS

In spite of the utility of a hindered-bed separator (i.e., classifier, density separator, desliming device), few innovations have been realized in recent years. However, because of the renewed interest by industry, several novel concepts have been developed and accepted. Specifically, a new hindered-bed separator has been developed that utilizes an innovative feed presentation system. A schematic of this device, which is known as the CrossFlow separator, is shown in Figure 2.

The CrossFlow utilizes a tangential, low-velocity feed entry system that introduces slurry at the top of the classifier. This approach allows water associated with the feed to travel across the top of the unit and report to the overflow launder with minimal disturbance of the separation chamber. To reduce the velocity of the feed flow, the feed stream enters a stilling well before flowing into the separation chamber. The feed then flows into the top of the device. Solids settle into the separation chamber as they travel between the feed entry point and overflow launder. The result of this feed presentation

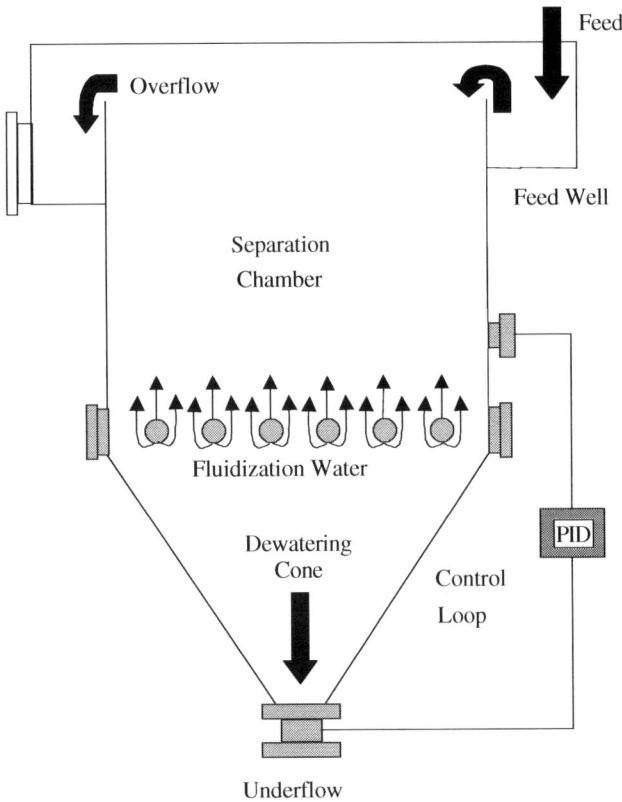

FIGURE 2 Schematic of crossflow hindered-bed separator

system, which eliminates excess feed water in the separation chamber, is an improvement in separation efficiency.

Past investigations (Dunn et al., 2000; Kohmuench et al., 2002) have verified that this novel feed presentation system allows for a more efficient separation as well as an increase in nominal capacity. These advantages are a result of a more quiescent separation chamber that is predominantly left undisturbed by any influx of water associated with the feed. This phenomenon is greatest in applications using little fluidization water. For instance, in a coal application treating 50 mtph (55 tph) at 50% solids, the influx of feed water is equivalent to 50 mtph (55 tph) or 220 gpm (13.9 L/s). This influx can have a large impact on a separator using a downcomer feed system. Essentially, the overall upward flow of water in greatly increased. In a 1.5 × 1.5 m (5 × 5 ft) separator, approximately 14.7 m^3/hr/m^2 (6 gpm/sqft) of teeter water is needed for fluidization; however, the water associated with the feed substantially increases the fluidization rate by 21.5 m^3/hr/m^2 (8.8 gpm/sqft) to 36.2 m^3/hr/m^2 (14.8 gpm/sqft). Consequently, the fluidization rate is more than doubled above the point of the feed entry. Using the new tangential feed presentation system, the entire separation chamber is left virtually undisturbed, allowing for a constant flow regime. This example is illustrated in Figure 3.

At a constant solids feed rate, lower feed percent solids increases the flow velocity in the upper portion of a conventional hydraulic classifier. This increase can misplace coarse (or dense) material to the overflow and leads to a decrease in efficiency (Heiskanen,

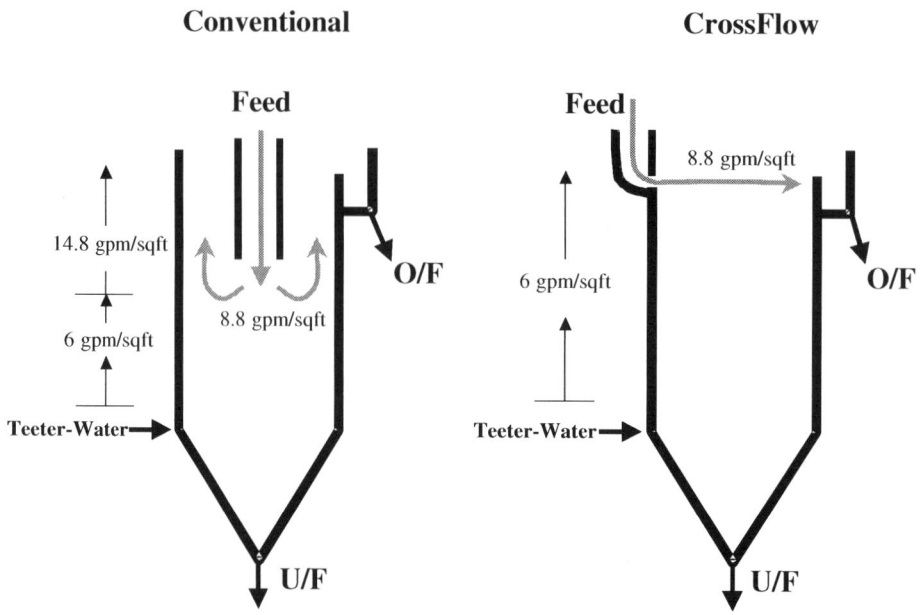

FIGURE 3 Comparison of conventional and CrossFlow separators

1993). Due to the nature of the feed system, however, the CrossFlow loses little coarse material to the overflow. The tangential feed system allows the excess feed water to flow across the top of the separation chamber, causing only a minimal amount of vertical flow disturbance.

Modeling investigations have verified the advantage of the tangential feed presentation system (Kohmuench et al., 2002). Presented in Figure 4 are the results of a modeling investigation in which the effect of feed percent solids (by mass) on process efficiency (Ep) was simulated at two different feed rates. At solid feed rates of 19.6 tph/m² (2.0 tph/ft²) and 36.2 tph/m² (3.7 tph/ft²), the feed percent solids were varied from a low of 20% to a high of 80%. All other process variables (i.e., teeter-water rate, cut-point, etc.) were held constant. These results indicate that the CrossFlow operates efficiently and with little variation over a wide range of volumetric feed flow rates. At the higher feed rate, the efficiency of the separation remained constant ($Ep \approx 0.120$) until the feed percent solids approached 35% by mass. At the lower feed rate, the efficiency of the separation ($Ep \approx 0.100$) remained consistent, even as the feed percent solids approached 25% by mass.

Further performance improvements can be realized when utilizing a lower dewatering cone (Figure 2). One advantage of the dewatering cone is that the underflow stream is discharged at a relatively high and constant percent solids. For instance, data from recent test work indicate that the percent solids of the underflow stream for a coal application (i.e., nominal 1 mm × 65 mesh) will discharge at 55% solids with little variation. This constant discharge dictates that the overall water split within the separator must remain constant.

In contrast, if an underflow is taken directly from the low percent solids teeter bed (i.e., flat bottom design), the tailings discharge will be highly variable and relatively wet. As the underflow valve opens and closes, the amount of water discharged will vary greatly and therefore lower the process efficiency. For example, as the underflow valve opens, more water is discharged directly from the teeter bed, which lowers the upward velocity of

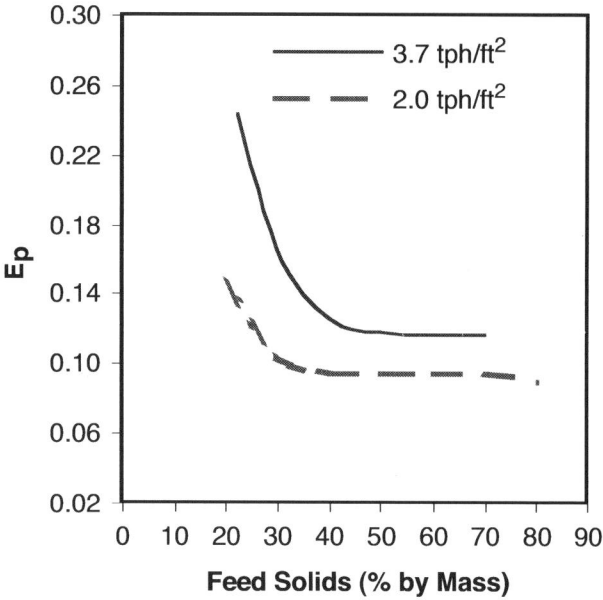

FIGURE 4 Modeling investigation results of tangential feed presentation system

teeter water, and hence the separation cut-point. This disadvantage can be overcome through the input of make-up water, which is added in proportion to the control valve position. This approach requires additional instrumentation and automatic control.

VERSUS SPIRALS

Teeter-bed separators continue to grow in popularity for the cleaning of fine coal (i.e., 2.0 × 0.250 mm). In particular, the popularity of these devices in Australia has far outweighed the interest of U.S. coal producers who primarily rely on single-stage coal spiral circuits. Spirals offer many advantages including high combustible recovery; however, coal spirals suffer from misplacement of coarse rock to the clean coal product. Additionally, coal spirals operate at a high specific gravity cut-point. Consequently, more efficient plant circuits must operate at reduced gravities to compensate for the high cut-point of the coal spirals.

Particularly detrimental to spiral performance is the misplacement of the rock and coal found between the clean coal and refuse splitters. This material, commonly referred to as middlings, is generally a mixture of misplaced coal and rock and not necessarily true middling material (i.e., half coal and half rock). Because of this phenomenon, the plant operator is faced with the choice of improving recovery while lowering product grade or improving product grade at the cost of recovery. Studies (Luttrell et al., 1998) have shown that separation efficiency and cut-point can be improved using two-stage spiral circuits that include middlings recycle. Regardless of the efficiency, these multistage spiral circuits have higher capital and operating costs when compared to single-stage spiral circuits and teeter-bed separators. In addition to the increased number of spiral starts, the multi-stage spirals also require the use of a pump and sump to handle the circulating middlings.

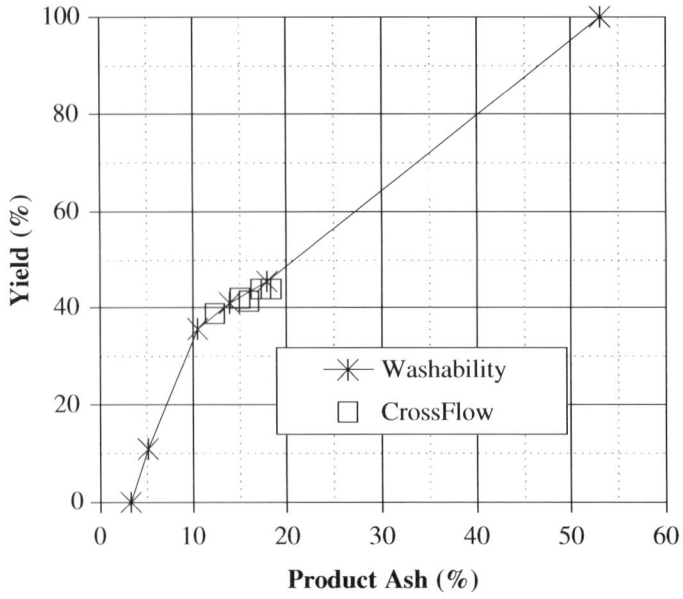

FIGURE 5 Laboratory-scale results versus washability

In contrast to the coal spiral, the teeter-bed separator offers many advantages. These advantages include a low separation cut-point (single stage), high-tonnage throughput per unit area of plant floor space, and automatic control of the separation cut-point. For example, a 91 mtph (100 tph) circuit for a general teeter-bed application would require a separator having approximately 4.5 m² (49 ft²) of cross-sectional area (20.2 mtph/m² or 2 tph/ft²). This can easily be accomplished in a unit measuring 2.1 × 2.1 meter (7 × 7 ft). In contrast, a coal spiral circuit would require at least a bank of 12 triple-start spirals that would consume close to 200 sqft (10 × 20 ft) of plant floor space. In addition, the required height of a hydraulic separator is less than or equal to the required height of the coal spirals when taking into account the spiral feed distribution system and product launders.

Furthermore, coal spirals are manually controlled (i.e., splitters), which make them extremely sensitive to changes in volumetric feed rate (Walsh and Kelly, 1992; Holland-Batt, 1994). Unlike coal spirals, the automatic control found on teeter-bed separators responds to plant fluctuations. As the solids loading increases or decreases due to a change in feed percent solids, the control system automatically compensates. In contrast, during these fluctuations the splitters on coal spirals must be manually adjusted as the coal and rock interface changes location on the spiral trough.

LABORATORY-SCALE TESTING

Eastern Australian Coal

Laboratory test work was conducted on an eastern Australian coal sample using the CrossFlow separator; the results are presented in Figure 5. During this test work, a washability analysis was conducted on the feed sample to provide a basis for determining efficiency. As shown, the CrossFlow hindered-bed separator was able to produce a low ash

TABLE 1 Efficiency results for laboratory testing

Results	CrossFlow Results		
	+28	28 × 100	Composite
Ep	0.073	0.105	0.108
Imp.	0.046	0.059	0.064
SG_{50}	1.569	1.758	1.669

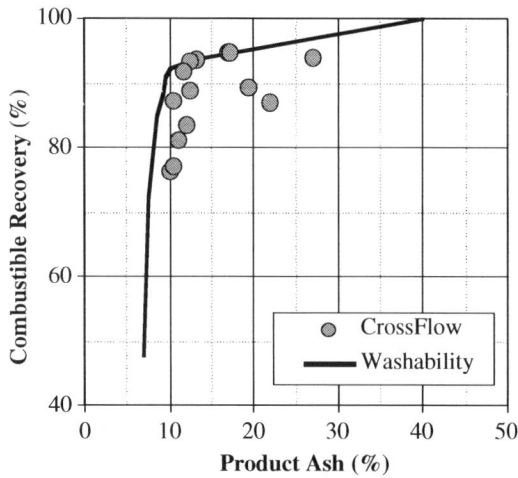

FIGURE 6 Pilot-scale results versus washability

product at a mass yield approaching the theoretical maximum. In fact, the organic efficiency for this test work approached 99%. Much of the coal lost to the tailings included the coarsest particles.

The separation efficiency data for the results shown in Figure 5 are presented in Table 1. The overall specific gravity cut-point for the laboratory testing was 1.67 SG. Separation efficiency for this testing was relatively high as indicated by the low Ep and imperfection numbers. As seen in this table, as the size fractions become coarser, the separation efficiency improves and the specific gravity cut-point is reduced. At finer sizes (i.e., –65 mesh), teeter-water flow overcomes the density of the particles and makes a separation predominantly based on size. Fortunately, material finer than 65 mesh is generally upgraded using flotation circuits that handle this size fraction very efficiently.

PILOT-SCALE TESTING

Central Appalachian Coal

Shown in Figure 6 are data produced using a pilot-scale CrossFlow separator. In this test work, feed was supplied to the CrossFlow directly from a coal spiral feed distributor. Feed percent solids were adjusted to approximately 35%. It can be seen that on this larger scale, the CrossFlow separator operates very close to the washability curve that is also presented in Figure 6. As presented in Table 2, at maximum separation efficiency, a product containing 12% ash was produced at a mass yield of 78% and combustible recovery of 92%.

TABLE 2 Metallurgical results for pilot-scale testing

Test (No.)	Product Ash (%)	O/F Mass Yield (%)	Combustible Recovery (%)
1	12.5	73.0	88.8
2	12.0	63.2	83.3
3	11.0	65.3	81.0
4	12.5	82.8	93.4
5	11.8	78.1	91.8
Avg:	12.0	72.5	87.7

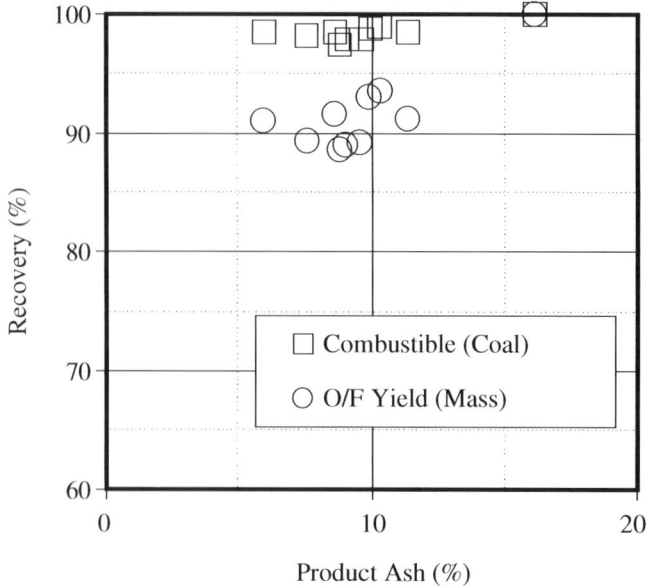

FIGURE 7 Pilot-scale results for low middlings coal

Further plant pilot-scale testing was completed on a coal with a low middlings content. As seen in Figure 7 and Table 3, the CrossFlow performed well, with an average combustible recovery of over 98%. Due to the relatively low feed ash of this coal, the average mass yield to the separator overflow was 90% with a mean product ash of 9%. The resultant tailings ash content was 85%.

FULL-SCALE INSTALLATION

Central Appalachian Coal

The pilot-scale testing of the low middlings central Appalachian coal led to the installation of a full-scale CrossFlow separator. This unit, 9 × 9 ft, was engineered to treat over 200 tph of coal. Performance results produced during commissioning are shown in Figure 8. The data indicate that the separation performance achieved at full scale is consistent with the performance demonstrated during the pilot-scale work. In fact, when optimized, this unit achieved a product ash content of 9.5% at a combustible recovery of 97.5%.

TABLE 3 Metallurgical results for pilot-scale testing of low midds coal

Test (No.)	Product Ash (%)	Reject Ash (%)	O/F Mass Yield (%)	Combustible Recovery (%)
1	9.1	84.4	89.00	97.9
2	11.4	85.9	91.2	98.5
3	10.0	85.5	92.9	98.8
4	8.7	85.1	91.6	98.5
5	5.6	85.6	91.1	98.5
6	7.6	85.4	89.3	98.1
7	10.3	86.8	93.4	99.0
8	9.7	83.9	89.1	97.9
9	8.8	81.8	88.5	97.5
Avg:	9.0	84.9	90.7	98.3

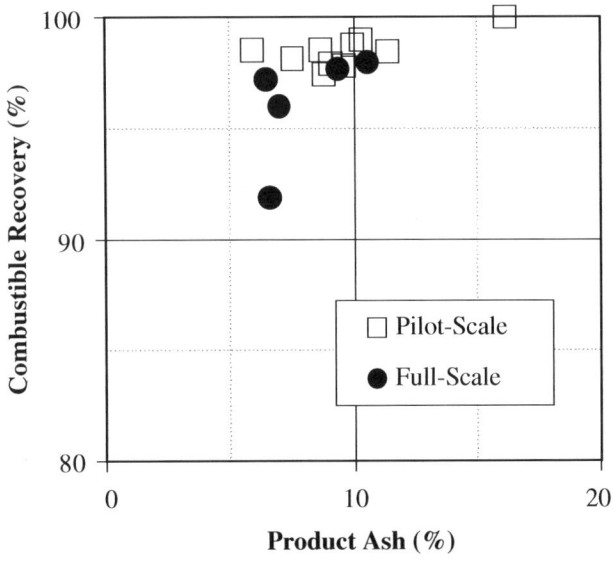

FIGURE 8 Full-scale commissioning results versus pilot-scale performance

SUMMARY AND CONCLUSIONS

1. A renewed interest by industry has been the incentive for several innovations in hindered-bed separator technology. The most recent of these innovations include a novel tangential feed presentation system that allows feed water to be immediately removed from the system.

2. Modeling investigations have shown that the CrossFlow, using the tangential feed presentation system, can maintain a high and less varied separation efficiency over a number of different operating conditions, including low feed percent solids (approaching 25% by mass).

3. Data show that teeter-bed separators are a good alternative to coal spiral circuits, especially single-stage circuits. These separators provide low specific gravity cutpoints, low capital and maintenance costs, and high solids throughput in comparison to coal spirals. More importantly, teeter-bed separators are less sensitive to changes in operating conditions such as changes in volumetric feed flow rate (i.e., percent solids).
4. Laboratory- and pilot-scale test data indicate that the CrossFlow separator can provide low ash products at high combustible recoveries. Data from this test work also show that the CrossFlow separator operates close to washability data providing high organic efficiencies.
5. A full-scale CrossFlow installation has shown that data collected during pilot-scale investigations are an acceptable indication of full-scale results (i.e., scale-up criteria). During commissioning, the full-scale separator achieved a combustible recovery of 97.5% and a product ash content under 10%.

REFERENCES

Dunn, P.L., Stewart, S.O., Kohmuench, J.N., and Cadena, C.A., 2000. "A Hydraulic Classifier Evaluation: Upgrading Heavy Mineral Concentrates," *Preprint 00-155*, SME Annual Meeting, February 28–March 1, 2000, Salt Lake City, Utah.

Heiskanen, K., 1993. *Particle Classification*, Powder Technology Series, Chapman and Hall, London, England.

Holland-Batt, A.B., 1994. "The Effect of Feed Rate on the Performance of Coal Spirals," *Coal Preparation*, Vol. 14, pp. 199–222.

Kohmuench, J.N., Mankosa, M.J., Luttrell, G.H., and Adel, G.T., 2002. "A Process Engineering Evaluation of the CrossFlow Separator," *Minerals and Metallurgical Processing*, February 2002, Vol. 19, No. 1, pp. 43–49.

Luttrell, G.H., Kohmuench, J.N., Stanley, F.L., and Trump, G.D., 1998. "Improving Spiral Performance Using Circuit Analysis," *Minerals and Metallurgical Processing*, November 1998, Vol. 15, No. 4, pp. 16–21.

Walsh, D.E., and Kelly, E.G., 1992. "An Investigation of the Performance of a Spiral Using Radioactive Gold Tracers," *Minerals and Metallurgical Processing*, Vol. 9, No. 3, pp. 105–109.

Innovations in Fine Coal Density Separations

R.Q. Honaker* and A.V. Ozsever*

A significant portion of the inefficiencies observed in today's coal preparation plants is related to the processing of −1 mm material. The improved liberation associated with the smallest particle sizes allows the opportunity for recovering greater amounts of higher quality material which, in turn, positively affects the entire plant operation. However, conventional technologies that are commonly employed to clean the fine coal are relatively inefficient, thereby preventing operators from fully realizing the benefits of fine coal recovery. Recently, several advances in fine coal density separators and associated circuitry have been developed and commercialized. This paper will review these advances and the impacts on overall plant performance.

INTRODUCTION

The reductions in separation density and improved process efficiencies associated with fine coal cleaning technologies have recently received significant attention. This trend is due to the realization of increased overall plant yield values and enhanced economical benefits (Luttrell et al., 1998; Luttrell et al., 1999; Honaker, Patwardhan, and Sevim, 1999). Additionally, expanding the applicability of density-based separators towards particle sizes below the typical 150-micrometer (100 mesh) is desirable due to the highly variable success of froth flotation processes for treating the ultrafine coal stream.

Conventional fine coal cleaning technologies such as single-pass spiral concentrator circuits typically provide relatively high separation density values of 1.8 and greater. Probable error values are around 0.15 to 0.20, which indicates moderate separation efficiency as compared to the values realized from processes cleaning the coarse coal fractions with values of 0.05 and lower (Osborne, 1988; Leonard and Hardinge, 1991).

Improving the fine coal cleaning efficiency is important for ensuring optimum overall plant yield while generating the required product grade. This fact is due to the ability to maximize the overall plant yield by maintaining constant incremental grades from

---

* Dept. of Mining Engineering. University of Kentucky, Lexington, Ky.

TABLE 1 Simulated separation performances showing the benefits of improved fine coal cleaning efficiencies and decreased separation densities when treating a run-of-mine coal containing 25% ash and 20% fine coal

Circuit	Case 1 Fine Circuit d_{50} = 1.8 Ep = 0.15		Case 2 Fine Circuit d_{50} = 1.6 Ep = 0.15		Case 3 Fine Circuit d_{50} = 1.6 Ep = 0.10	
	Mass Yield %	Product Ash %	Mass Yield %	Product Ash %	Mass Yield %	Product Ash %
Simulation 1						
Coarse	24.9	3.94	34.6	4.82	36.3	5.03
Fine	14.8	10.04	13.1	9.07	13.9	8.55
Plant	39.7	6.22	47.7	6.00	50.2	6.00
Simulation 2						
Coarse	42.4	5.94	45.2	6.40	45.9	6.54
Fine	14.8	10.04	13.1	9.07	13.9	8.55
Plant	57.2	7.00	58.3	7.00	59.8	7.00

each process circuit (Abbot, 1981; Salama, 1986). A coal preparation plant typically utilizes two to four coal cleaning circuits to treat the wide range of particle size fractions that exist in run-of-mine coal. Thus, to achieve maximum plant yield, the quality of the last mass increment (i.e., "dirtiest" particles) recovered from each circuit must be equal. This implies the need for equivalent separation density values in each circuit.

To quantify the typical separation performance improvement, simulations were conducted using a modified Whiten equation to generate the partition curve values. The run-of-mine coal data was obtained from an eastern U. S. coal operation. Approximately 80% of the coal had a particle size greater than 1 mm (16 mesh), for which 14.7% by weight had a relative solid density between 1.6 and 2.0. The same density fraction in the –1 mm material represented 4.4% of the total weight. A dense-medium process providing a 0.03 Ep value was assumed for treating the +1 mm material while the separating density was varied to achieve the required product grade. Three scenarios were considered for cleaning the fine –1 mm material with each resulting in a 10% by-pass of heavy, high-ash particles to the product stream. Case 1 represented the typical cleaning performance with a relative separation density of 1.8 and an Ep of 0.15. The second case considers a reduction in separation density to 1.6 while the third involves Case 2 and an improvement in the Ep to 0.10.

When considering the production of a required 6% product ash, the typical operating conditions represented by Case 1 were unable to completely satisfy the desired grade as shown in Table 1. This is due to the significant loss in yield as the separation density approaches 1.30 in the coarse cleaning circuit. However, by reducing the separation density achieved by the fine coal cleaning circuit from 1.80 to 1.60, the 6.00% ash product is realized while increasing the plant yield by 8 absolute percentage points beyond the best Case 1 result. Improving the fine circuit efficiency as indicated by a drop in the Ep value to 0.10 allows an additional 2.5 absolute percentage point gain and raises the required separation density to 1.41 for the coarse circuit. The less stringent requirement of 7% product ash content reduces the overall impact of the fine circuit improvements. However, the plant yield enhancements are significant at 1.1% for Case 2 and a cumulative 2.6% for Case 3 while the coarse coal separation density is more desirable with values exceeding 1.50.

The economical benefits realized from the improved separation performance are substantial. Consider a typical processing plant that treats 1,000 tonnes/hr of coal and operates 4,000 hours annually. For every one absolute percentage point increase in plant yield, the annual clean coal production increases by 40,000 tons. This equates to an average revenue gain of around $1 million U.S. and a corresponding reduction in mining and cleaning costs on a clean coal basis. As shown in Table 1, improving the fine coal cleaning performance has the potential to significantly enhance plant yield by several percentage points. The overall impact is obviously a function of the required product grade, the liberation characteristics of the coal, and the separation processes utilized in the circuit. However, the demonstrated potential explains the interest in improving fine coal cleaning.

Froth flotation is a process commonly employed in industry to treat the −150 micrometer (−100 mesh) particle size fraction. The process is sensitive to operating conditions and the interactive parameter effects complicate the basic understanding of the process. As a result, significant effort has been directed toward modifying conventional processes and developing new technologies that will increase the applicability of density-based separators for treating ultrafine particles. Devices that apply a centrifugal field by mechanical action to conventional cleaning methods such as jigging have shown to provide effective fine coal separations (Riley, Firth, and Lockhart, 1995; Luttrell, Honaker, and Philips, 1995; Honaker, Mohanty, and Govindarajan, 1998).

This paper will review the developments associated with fine coal density separations and discuss the potential application of each technology. Improvements in the overall separation performance will be provided and discussed in accordance with the impact on the overall plant.

SPIRAL CONCENTRATOR CIRCUITS

Spiral concentrators provide a simple, low cost separation that requires minimal operator attention and generally provides exceptional coal recovery. Due to these important attributes, spirals account for an estimated 9% of the total worldwide coal processing capacity and 12% of the U.S. capacity (Kempnich, 2000). However, subsequent to the significant number of spiral concentrator installations in the 1980s, attention was directed towards improving the noted deficiencies of the unit. The performance deficiencies include a relatively high separation density of around 1.8 for rougher-only applications, an average process efficiency as represented by probable error (Ep) values in the range of 0.15–0.18 and an inherent by-pass of coarse high density particles in the product steam. In a study reported by MacNamara et al. (1995), approximately 13.5% of the clean coal product from a spiral circuit treating 1 × 0.15 mm coal had a relative density greater than 1.7. In addition, on-line separation control is difficult for spiral units.

The typical spiral concentrator used for cleaning fine coal produces three process streams, i.e., product, middlings and tailings (refuse) streams. In rougher-only applications, the middlings material is typically combined with the tailings stream and thus the product quality and yield is based on the splitter position between the product and middling flows. Performance evaluations conducted by several plant operators and researchers found that a significant amount of clean coal existed in the middlings stream, which was being misplaced to the overall fine coal refuse.

In an effort to alleviate the inefficiencies associated with the middlings stream, two-stage spiral circuits were proposed by Richards, Hinter, and Holland-Batt (1985). A two-stage middlings retreatment circuit was installed in a U.S. processing plant in 1988. As shown in Figure 1, the middlings stream from the primary spiral is retreated in a secondary spiral. The product of the secondary spiral is combined with the primary

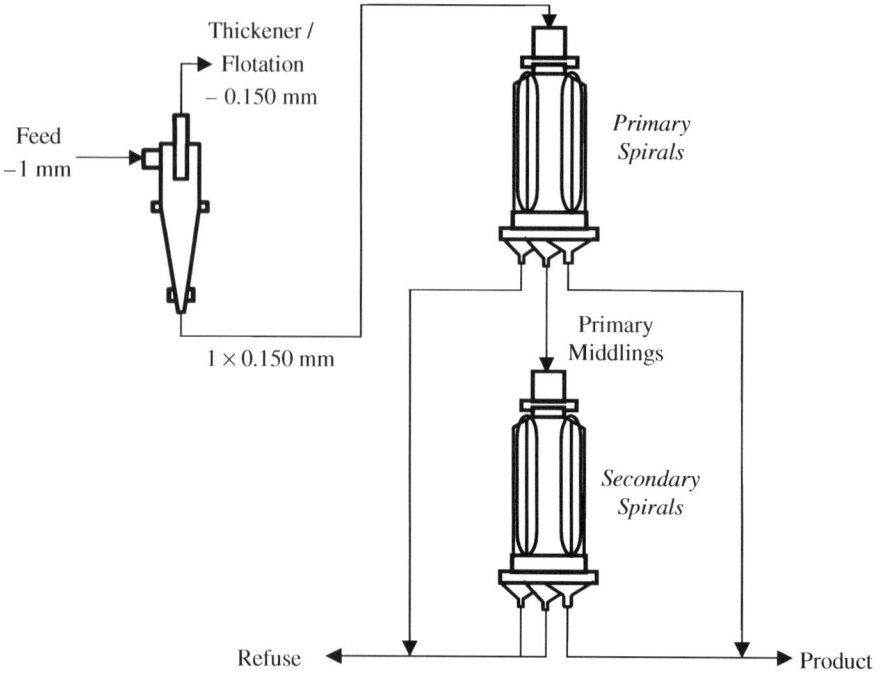

FIGURE 1 Flowsheet of a two stage, middlings retreatment spiral circuit without the use of a recycle stream

spiral product to provide the overall 1 × 0.15 mm clean coal concentrate. The middlings material from the secondary spiral is combined with the two tailing streams and disposed as reject. According to Bethell, Stanley, and Horton (1991), the two-stage circuit decreased the Ep value achieved from cleaning the fine coal from around 0.15 to 0.135, which enhanced organic efficiency by five absolute percentage points. The separation density remained relatively unchanged at around 1.74. This efficiency improvement was found to benefit the overall plant performance when the separation densities utilized in the coarser process circuits were below 1.7. Under typical operating conditions, an annual increase in clean coal production of 6,600 tonnes/hr was estimated as a result of the second stage middlings treatment, which enhanced the annual net revenue by about $165,000 US.

As a result of the two-stage middling retreatment benefits, developmental work was initiated to develop a single spiral that incorporates the concepts of the two-stage process. This effort utilized the general findings that indicated the potential to reduce the number of turns on a spiral from the typical 5 to 7 turns to less than 4 turns. As a result, single-stage spirals would require less headroom. Also, a middlings retreatment portion could be added on the same axis without significantly increasing the overall length beyond the conventional spiral. MacNamara et al. (1995) reported that no differences in separation performances were observed in tests conducted on a 3-¼ turn spiral and a 5-¾ turn spiral. Based on these findings, a new "compound" spiral was developed which utilizes the first 2-¼ turns as a primary treatment stage and the final 2 turns to treat the middlings from the primary stage. In plant tests involving a 1.18 × 0.250 mm particle size fraction revealed a relative separation density of 1.59 with an Ep value of 0.23 in the presence of about 11% near-gravity material (MacNamara et al.,

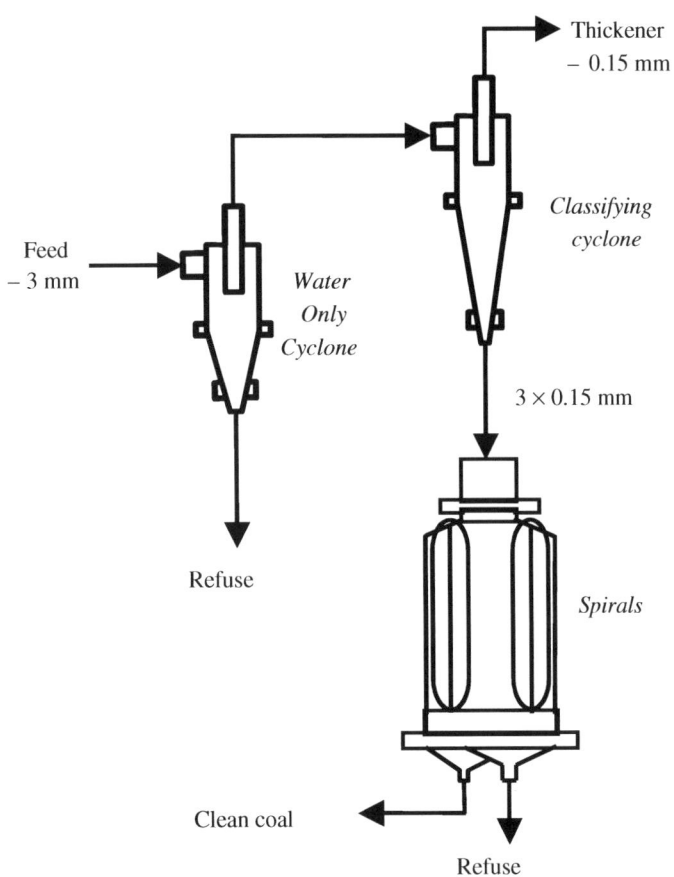

FIGURE 2 Flowsheet of a WOC—spiral circuit

1996). The separation density was significantly reduced from the typical values but the separation efficiency depreciated, which is dissimilar to the findings from the separate two-stage spiral results (Bethel et al., 1991).

The reduced process efficiencies that were observed from the middlings retreatment spiral might be associated with the spiral length effects. Although MacNamara et al. (1995) reported minimal differences in performance, Atasoy and Spottiswood (1995) reported that separation density and the by-pass of high-density material reduces with increasing spiral length. A 5 to 6 turn spiral was recommended to achieve optimum fine coal separations. In a separate study, Weldon and MacHunter (1998) conclude that 4 turn spirals is optimum for most applications and that a conventional 5-¼ turn spiral is required to achieve low-density separations.

As described previously, the by-pass of coarse, high-density material in the product stream is an inherent problem associated with spiral concentrators. Approximately 10% high-density by-pass has been reported when considering only the 1 × 0.15 mm fraction while amounts as high as 18% occurred for the +1 mm fractions (MacNamara et al., 1995; Atasoy and Spottiswood, 1995). In an effort to remove the coarse, high-density material, a circuit has been employed that utilizes a water-only cyclone (WOC) as pretreatment to the spiral concentrators. As shown in Figure 2, the WOC treats the desliming screen underflow. The WOC overflow is classified using cyclones to achieve a 0.15-mm particle size separation. The underflow of the classifying cyclone serves as feed to the spiral concentrator. MacNamara et al. (1995) reported on an in-plant study of this circuit for the treatment of 3 × 0.15 mm coal. The WOC-spiral circuit reduced the cut-point of the 3 × 1 mm size fraction from 1.77 to 1.61 while maintaining the process efficiency at a near constant value. This positive effect allowed the densities in the coarse medium vessel to be increased by five to eight points, thereby resulting in a substantial increase in plant yield. However, an operational concern with this circuit is the potential loss of coarse coal to the underflow of the WOC, especially as the apex experiences wear with an increase in operating time.

A more fundamental analysis of various spiral circuit arrangements was performed by Luttrell et al. (1998) using the concept of linearity. Various arrangements involving rougher, scavenger and cleaner spirals were considered to identify the circuit with the greatest potential for achieving optimum process efficiency and low separation densities while maintaining operational simplicity. In general, any circuit that provided a feedback stream to a previous unit provided better performance than a single unit. For example, the circuit that recirculates the product from the second stage scavenger unit to the feed of the rougher has 1.33 times greater efficiency than a single stage process. Interestingly, the middlings retreatment circuit described in Figure 1 was found to provide equal efficiency to that obtained from a single stage rougher spiral.

Based on the separation efficiency evaluation and operational considerations, a circuit utilizing a rougher-cleaner approach was selected for installation into an operating coal preparation plant. As shown in Figure 3, the cleaner spiral treats the product and middlings material from the rougher spiral. The middlings stream from the cleaner spiral is recycled to the feed of the rougher unit. The tailings from the two units are combined for total refuse.

The rougher-cleaner circuit with recycle achieved a relative separation density of 1.65 (Luttrell et al., 1998). In comparison to a rougher-cleaner circuit that does not utilize middling recycle, the separation density was nearly equal. However, separation efficiency for the circuit utilizing recycle was found to be 1.2 times (0.15 versus 0.18 Ep) that achievable by a circuit not using recycle. The result was a plant yield increase of 3.87%, which equated to an estimated annual revenue increase of $255,000.

TEETER-BED SEPARATORS

Teeter-bed separators (TBS) have been employed for particle size classification since the early part of the twentieth century in the mineral and coal processing industries. However, recent developments in automation and control have provided an opportunity to fine-tune the operating conditions of TBS units so that effective density separations can be achieved for fine particles. As shown in Figure 4, teeter-bed units utilize an upward current to suspend high-density particles entering in the feed stream. The result is a hindered-settling environment that provides an autogenous dense medium-type separation. Low-density coal is unable to settle through the high-density particle bed and is transported by the upward movement of the teeter water into the overflow process stream. High-density, ash-bearing particles are able to penetrate the bed and settle into the underflow stream. Pressure transducers and underflow valves linked to a controller adjust and maintain the fluidized particle bed at the desired level.

Mankosa, Stanley, and Honaker (1995) conducted tests using a teeter-bed classifier as a pre-cleaner to a spiral concentrator bank for the treatment of a difficult-to-clean fine coal. The findings suggested that the teeter-bed–spiral combination provides an efficient

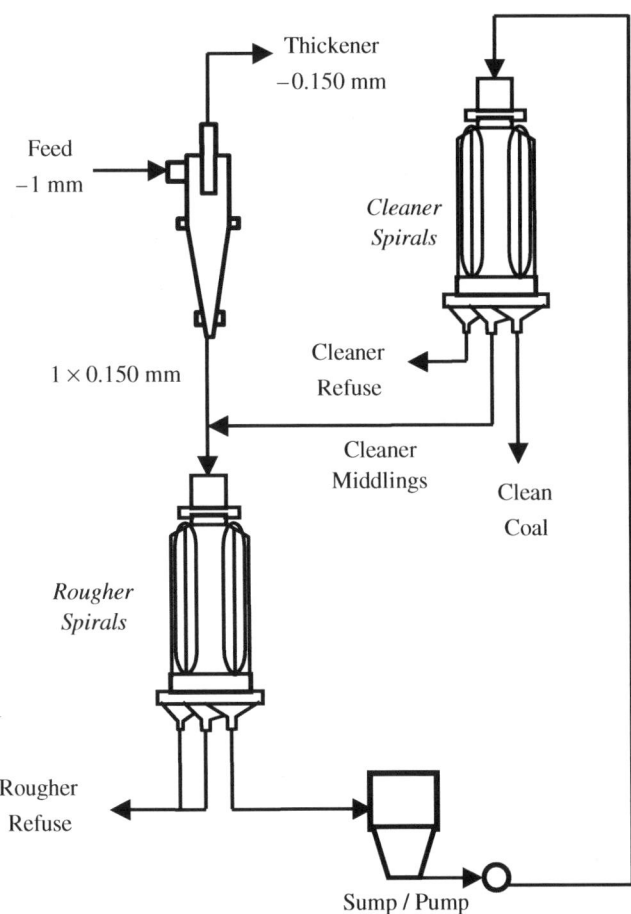

FIGURE 3 Flowsheet of a rougher-cleaner spiral circuit in which the product and middlings streams of the rougher spiral bank is retreated in a cleaner unit with middlings recycle

separation as indicated by an Ep value of 0.07 for a 1 × 0.15 mm Illinois No. 5 coal. The relative separation density was approximately 1.8. Studies comparing the separation performances achieved by spiral concentrators and TBS units found the TBS to provide more efficient gravity separations for nominally 1 mm × 150 μm coal (Reed et al., 1995; Honaker, 1996). Nicol and Drummond (1997) have described the efficiency of TBS units and several fine coal circuits for which application may prove to be beneficial. Based on the acceptance of the preliminary studies, a 75 tonnes/hr TBS unit was installed in an operating coal preparation plant in which the separator is being used as a cleaner unit for the spiral product (Drummond et al., 1998). This application reportedly provides relative separation densities (d_{50}) below 1.6 and probable error (Ep) values in the range of 0.10 to 0.15.

A flow diagram of a typical circuit that employs the use of a TBS unit for fine coal cleaning is shown Figure 5. The TBS unit is utilized as the primary fine coal-cleaning unit whereby the 1 × 0.15 mm feed is provided from the underflow of a classifying cyclone bank. A clean coal product is obtained from screening the overflow at 0.3 mm. The screen overflow is a clean concentrate while the underflow is processed through a spiral bank to

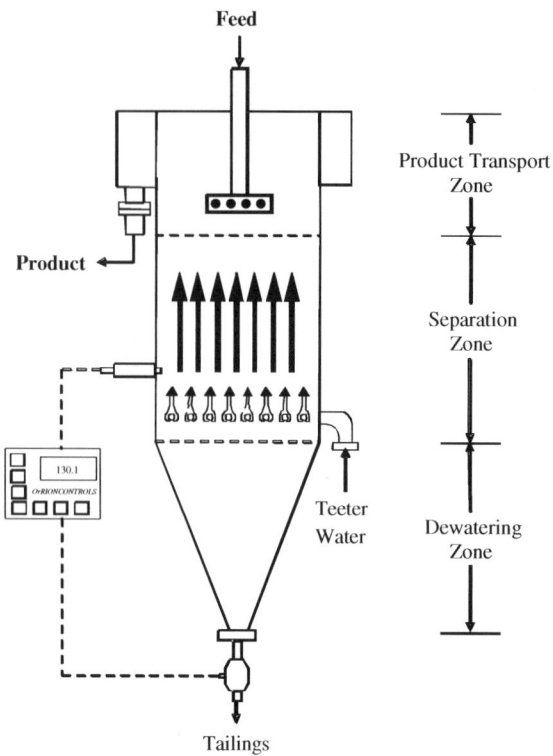

FIGURE 4 Schematic of a teeter-bed separator with an indication of specific action zones

generate a second clean coal stream. Utilizing a similar circuit to treat 2 × 0.25 mm coal, an Australian preparation plant achieved a low relative separation density of 1.49 while achieving an Ep value of 0.094 (Drummond, Nicol, and Swanson, 2002).

Tests have also been conducted on flotation tailing streams at another Australian processing plant where the feed is comprised of –1 mm material. The benefit to the TBS is that the clay slimes in the flotation tailings provide an increase in the medium density. From these tests, the TBS unit obtained separation densities below 1.45 while realizing exceptional process efficiencies with Ep values below 0.05 (Drummond, Nicol, and Swanson, 2002).

Expanding the particle size application range for TBS units to 3 × 0.15 mm has been realized by incorporating a two-stage process in one unit. The separator incorporates an internal de-shaling step to remove the coarse gangue material followed by treatment in an outer teeter bed (Snoby, Grotjohann, and Jungmann, 1999). From the treatment of a 2.4 × 0.15 mm North American coal, the ash content was reduced from 34.2% to 8.5% for which the Ep value was found to be approximately 0.12.

In a further effort to expand the effective particle size range, a commercial TBS unit has been designed with bubble injection into the teeter water (Mankosa, 1999). The air bubbles collide and attach to the coarse coal particles thereby decreasing the effective density. The low-density coal-bubble aggregate is effectively elutriated to the top of the separator along with the fine coal and ultrafine gangue material. Classification of the overflow stream results in a final clean coal product.

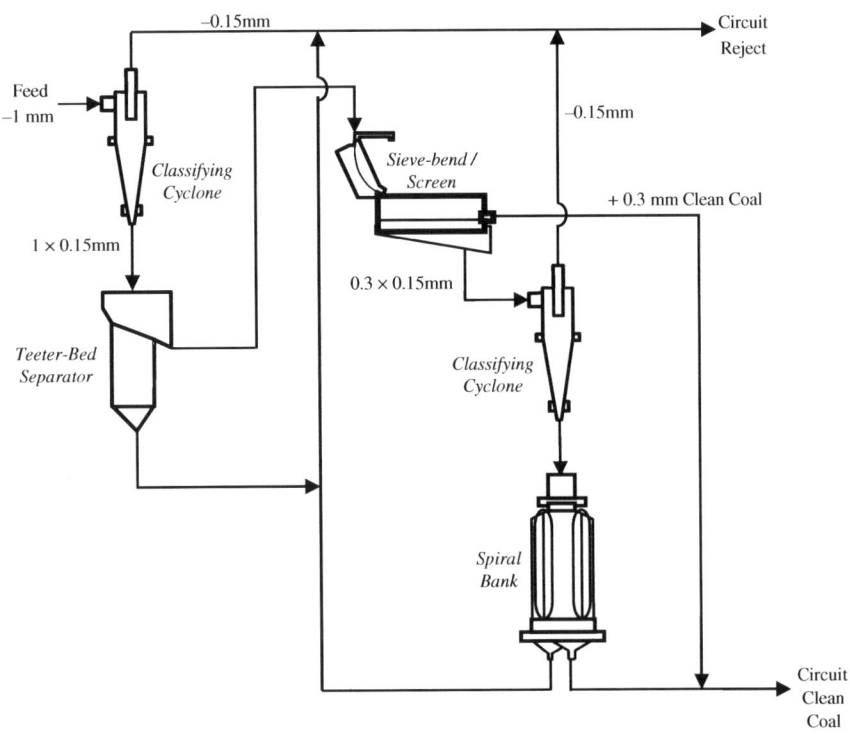

FIGURE 5 Flowsheet showing the application of a teeter-bed separator for fine coal cleaning

ENHANCED GRAVITY SEPARATORS

A significant need exists for reducing the lower particle size limitation associated with density-based separations, which is about 0.15 mm. Froth flotation is the most commonly used process for treating the –0.15 mm particle size material as indicated by a recent survey that states the process represents 14% of the worldwide installed plant capacity (Kempnich, 2000). However, several factors favor the increased use of density separations for treating ultrafine coal as an alternative or as an enhancement to froth flotation, i.e.,

1. Coals sometimes vary in their degree of hydrophobicity within the seam. As a result, poor flotation recovery is periodically realized. The variability of hydrophobicity and other surface chemistry factors is significantly greater than the variations in particle density distribution. Thus, density-based separations would provide a more effective and stable performance.

2. Coal pyrite typically concentrates in the froth product due to natural surface hydrophobicity. Due to the large density difference between coal (sp.gr. = 1.3) and pyrite (sp.gr. = 4.8), physical sulfur rejection is better achieved using density-based separations.

3. Density-based separations are fundamentally more efficient than the separations achieved by froth flotation systems. A particle comprised of only 10% coal may float due to bubble-particle attached at the coal-air interface. However, the high particle

density would result in rejection to the tailings stream. Fundamental studies have affirmed this hypothesis (Drum and Hogg, 1987; Adel, Wang, and Yoon, 1989; Killmeyer, Hucko, and Jacobsen, 1992; Mohanty, Honaker, and Ho, 1998).

4. The capacity of flotation systems is significantly lower on the basis of required floor space as compared to density-based systems. The typical capacity of a flotation system for –0.15 mm coal is around 1.3 tonnes/hr/m^2; whereas density-based systems provide 5 tonnes/hr/m^2 or greater depending on the mechanisms employed.

5. Typically, froth flotation principles are not well understood by coal plant operators. As a result, the systems often run under non-optimum conditions. On the other hand, the concepts of density-based separations are easier to comprehend and manipulate in an effort to achieve improved performance.

With the goal of providing efficient density-based separations for ultrafine coal, enhanced gravity separators (EGS) were developed and commercialized in the 1990s. The EGS units utilize conventional density-based separation principles and an enhanced gravity field provided by a mechanical spinning action. Descriptions of three different EGS units are provided in the following paragraphs.

Centrifugal Jigs

The typical mechanisms of a centrifugal jig are schematically shown in Figure 6(a). The feed enters into a spinning basket that is placed inside a static casing having launders for product and tailing collection. Water enters into the hutch either through a control valve or, as shown in Figure 6(a), through a pulsation block that cycles with the spinning bowl. During the entire process, the tailings material, which passes through the basket and into the hutch, is discharged through a series of valves that are equally spaced along the outer circumference of the static casing. The clean coal product overflows a lip at the end of the cage, which is adjustable to change the particle bed depth. Two commercial units include the Kelsey (Geo Logics Pty. Ltd., Lonsdale, Australia) and Altair (Mineral Recovery Systems Inc., Reno, NV) centrifugal jigs.

Flowing Film

Figure 6(b) shows an EGS unit that utilizes flowing-film principles similar to launders and spiral concentrators. Feed slurry enters vertically downward into the open face of a spinning solid bowl. A rotor at the bottom of the bowl accelerates the particles toward the bowl wall and initiates stratification process based on particle density differences. The solids are transported up the bowl wall by a component of the applied centrifugal force. The high-density particles along the bowl wall fall into a slot that exists along the entire circumference of the upper portion of the bowl. Control valves placed within the slot are manipulated by air pressure to obtain the desire tailings flow rate. The low-density material overflows a lip and is discharged to the overflow collection launder. The Falcon Concentrator (Falcon Concentrators, Langley, B.C., Canada) is a commercially available technology utilizing a smooth bowl while the Multi-Gravity separator (Axsia Mozley, Cornwall, UK) employs flowing-film principles as related to riffle tables.

Teeter Bed

One can imagine a teeter-bed separator such as the unit depicted in Figure 4 that is suspended by a support system and spun around along a central axis. The resulting centrifugal force accelerates the process described in the previous section and thus provides the

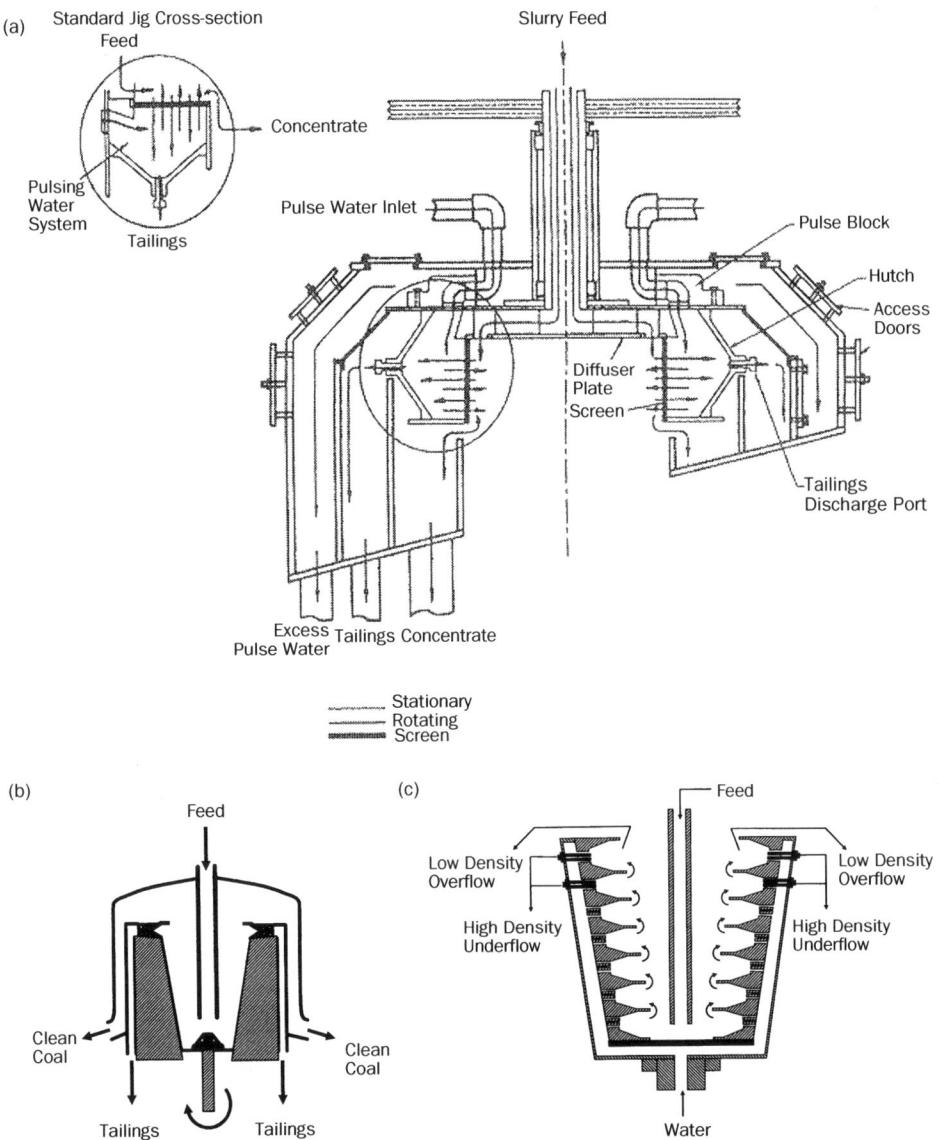

FIGURE 6 Schematic diagrams of centrifugal (a) jigging, (b) flowing-film, and (c) teeter-bed separators

potential of achieving similar process performances on ultrafine particles. Figure 6(c) shows a diagram of a centrifugal teeter-bed separator whereby the teeter water enters through tangential holes placed inside vanes that spiral toward the top of a spinning bowl. The Knelson Concentrator (Knelson Concentrators, Langley, B.C., Canada) is the only known commercial unit utilizing this concept.

Studies have shown that the EGS units have the ability to provide separation efficiencies and densities for ultrafine coal cleaning that are comparable to those achieved by their respective conventional systems used for coarse coal cleaning. Table 2 lists the typical

TABLE 2 Typical probable error values and separation densities achieved by various enhanced gravity separator types for cleaning coal in the 210 × 37 μm fraction

Parameters	Flowing Film	Teeter Bed	Jigging
Max Centrifugal Force	300	60	60
Separation Density	1.5–1.8	1.9	1.44–2.1
Probable Error	0.10–0.15	0.10	0.10–0.12
References	Venkatraman (1995) Honaker (1998)	Honaker et al. (1998)	Riley et al. (1995) Mohanty et al. (1999)

probable errors and specific cut points achieved by various EGS types. An evaluation of each technology has been reported by Luttrell, Honaker, and Philips (1995) and Honaker, Mohanty, and Govindarajan (1998).

In a recent study, the effect of applying a dense medium to a centrifugal flowing-film concentrator was conducted. Honaker, Singh, and Govindarajan (2000) reported that, for a difficult-to-clean coal, the mass yield improvement compared to a water-only system was 18 absolute percentage points at a given product grade. Separation density values as low as 1.40 were achieved while realizing Ep values below 0.05. An optimization study reported that the dense medium EGS application achieved a 95% organic efficiency when treating 1 × 0.45 mm coal (Honaker and Patil, 2002).

An application for EGS units in fine coal circuits is the treatment of the nominal 150 × 44 μm size fraction as shown in Figure 7. The –1 mm circuit feed is initially classified to achieve a 150 μm size separation using a primary classifying cyclone bank. The nominal 1 × 0.15 mm cyclone underflow is treated in spiral concentrators. The primary classifying cyclone overflow stream is reprocessed through a secondary cyclone bank to achieve a 44 μm size separation. The –44 μm size fraction often contains mostly submicron clay and existing coal is difficult to recover and dewater. The 150 × 44 μm cyclone underflow is fed to the EGS unit to clean and recover the ultrafine coal. Since a significant amount of clay is typically by-passed to the classifying cyclone underflow, the EGS product stream is processed through a screen-bowl centrifuge to dewater and provide additional rejection of the submicron material, which typically amounts to about 50% of the total submicron material in the feed. This application represents a direct substitution for froth flotation. The circuit is commonly employed in U.S. preparation plants with froth flotation systems to treat the de-slimed 150 × 44 μm material, especially in operations employing conventional flotation systems.

In a previous study, an enhanced gravity unit was employed as a cleaner unit for froth flotation concentrate (Venkatraman, Luttrell, and Yoon, 1995). The application provided a significant benefit for the rejection of pyritic sulfur, which tends to be concentrated in the froth product. Honaker, Singh, and Govindarajan (2000) found that the separation efficiency achieved on the 150 × 44 μm fraction by an EGS unit was more efficient than that achieved by froth flotation. Since it is fundamentally better to locate the most efficient unit as the primary unit, it is suggested to employ the EGS as a rougher unit with flotation serving as the cleaner process for the EGS product stream as shown in Figure 8. An additional advantage of this arrangement is the presence of a significant amount of clay particles in the EGS unit, which tends to increase the density of the medium thereby improving process efficiency. A concern with this application is the loss of coarse coal to the tailings stream of the EGS, which is more prominent for particle sizes greater than 150 μm.

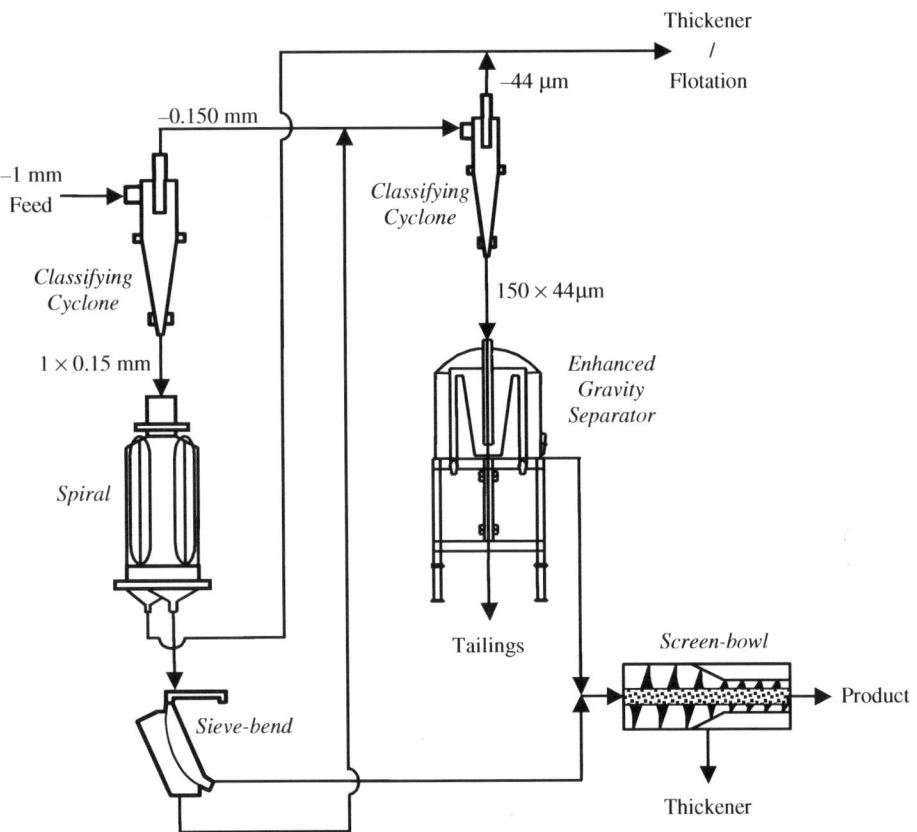

FIGURE 7 Flowsheet showing the application of enhanced gravity separators for cleaning coal in the 150 × 44 μm particle size fraction

SUMMARY AND CONCLUSIONS

The recent developments in fine coal density separations generally address (1) the development of innovative circuitry arrangements using conventional cleaning technologies to reduce separation density and/or process efficiency or (2) the development of novel technologies to allow effective density-based separations for particle sizes less than 150 μm. Reducing separation density below the typical values achieved for −1 mm coal improves the opportunity for preparation plants to operate under constant incremental quality conditions. As a result, the maximum plant yield can be realized while producing a given product quality, which significantly improves annual revenue and profit potential.

Research involving fundamental evaluations of various fine coal circuits involving spiral concentrators and in-plant testing of the circuits has revealed the ability to achieve relative separation densities of approximately 1.6 with an Ep of around 0.15. This finding typically improves the overall plant yield by 1 to 2 percentage points depending on the characteristics of the plant feed and the coarse coal cleaning circuit(s). In-plant installations incorporating teeter-bed separators and spiral concentrators have observed similar separation densities with a reduction in the Ep to around 0.10. Fine coal circuits providing a separation density of 1.6 and an Ep of 0.10 have the

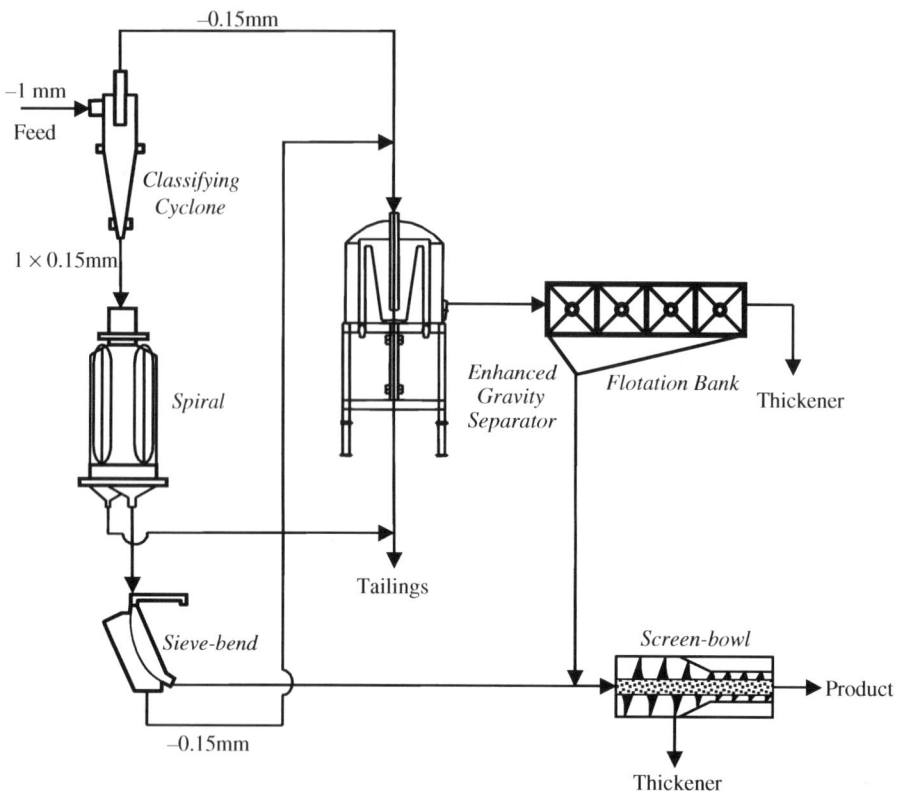

FIGURE 8 Flowsheet showing the application of enhanced gravity separators as a pre-cleaner to froth flotation

potential to improve overall plant yield by 2 percentage points or more based on incremental quality optimization.

The development of enhanced gravity separators (EGS) provides the opportunity to achieve effective density separations for −150 μm material. As a result, EGS units may recover ultrafine coal that fails to respond satisfactorily to froth flotation processes. Fully continuous EGS units are commercially available with capacities that approach 100 tonnes/hr. Although centrifugal separators are used extensively in the minerals industries, the installation of a unit in an operating coal preparation plant has not been realized. This fact is mainly due to the continuing development of the technology and the concern with the inability to remove the slime material from the product stream without the need for froth flotation.

REFERENCES

Abbot, J. 1981. The optimization of process parameters to maximize the profitability from a three component blend. *Proceedings First Australian Coal Preparation Conference.* Newcastle, Australia. 87–106.

Adel, G.T., Wang, D., and Yoon, R.H. 1989. Image analysis characterization of coal flotation products. In *Advances in Coal and Minerals Processing Using Flotation*, Chapter 5, Edited by S. Chander, Littleton, CO: Society for Mining, Metallurgy, and Exploration, Inc. 45–53.

Atasoy, Y., Spottiswood, D.J. 1995. A study of particle separation in a spiral concentrator. *Minerals Engineering*. 8(10): 1197–1208.

Bethell, P.J., Stanley, F.L., and Horton, S. 1991. Benefits associated with two-stage spiral cleaning at McClure river preparation plant. *Minerals & Metallurgical Processing*. 8(4): 215–219.

Dumm, T.F., Hogg, R. 1987. Washability of ultrafine coal. *Proceedings SME Annual Meeting*, Denver, CO, Preprint no. 87-136.

Drummond, R.B., Swanson, A.R., Nicol, S.K., and Newling, P.G. 1998. Optimization studies on a 75 t/h teetered bed separator at Stratford coal. In *XIII International Coal Preparation Congress and Exhibition*. Vol. 1. Edited by Partridge, A.C. and Partridge, I.R. Newcastle, Australia: South African Institute of Mining and Metallurgy. 215–224.

Drummond, R.B., Nicol, S., and Swanson, A. 2002. Teeter bed separators–the Australian experience. In *XIV International Coal Preparation Congress and Exhibition*. Sandton, South Africa: South African Institute of Mining and Metallurgy. 353–358.

Honaker, R.Q. 1996. Hindered-bed classifiers for fine coal cleaning. *Proceedings 13th International Coal Preparation Conference*. Intertec Inc., Lexington, KY. 59–70.

Honaker, R.Q., Mohanty, M.K., and Govindarajan, B. 1998. Enhanced gravity concentration: An effective tool for fine coal cleaning, In *Proceedings of the XIII International Coal Preparation Congress*. Edited by Partridge, A.C. and Partridge, I.R. Newcastle, Australia: Australian Coal Preparation Society. 349–357.

Honaker, R.Q., Patwardhan, A., and Sevim, H. 1999. Improving mine profitability through advanced fine coal cleaning. *Proceedings 16th International Coal Preparation Conference*. Intertec Inc., Lexington, KY. 106–123.

Honaker, R.Q., Singh, N., and Govindarajan, B. 2000. Application of dense-medium in an enhanced gravity separator for fine coal cleaning. *Minerals Engineering*. 13(4): 415–427.

Honaker, R.Q., Patil, D.P. 2002. Parametric evaluation of a dense-medium process using an enhanced gravity separator. *Coal Preparation*. 22: 1–17.

Kempnich, R. 2000. Coal Preparation–A world view. *Proceedings 17th International Coal Preparation Conference*. Intertec Inc., Lexington, KY. 5–48.

Killmeyer, R.P., Hucko, R.E., and Jacobsen, P.S. 1992. Centrifugal float-sink testing of fine coal: An interlaboratory test program. *Coal Preparation*. 10: 107–118.

Leonard, W., Hardinge, B.C. 1991. *Coal Preparation*. Society for Mining, Metallurgy and Exploration, Inc. Littleton, CO.

Luttrell, G.H., Honaker, R.Q., and Philips, D. 1995. Enhanced gravity separators: New alternatives for fine coal cleaning, *Proceedings 12th International Coal Preparation Conference*, Intertec Inc., Lexington, KY. 282–292.

Luttrell, G.H., Kohmuench, J.N., Stanley, F.L., and Trump, G.D. 1998. Improving spiral performance using circuit analysis. *Minerals & Metallurgical Processing*. 15(4): 16–21.

Luttrell, G.H., Kohmuench, J.N., Stanley, F.L., and Trump, G.D. 1999. An evaluation of multi-stage spiral circuits. *Proceedings 16th International Coal Preparation Conference*, Intertec Inc., Lexington, KY. 80–88.

MacNamara, L., Addison, F., Miles, N.J., Bethell, P., and Davis, P. 1995. The application of new configurations of coal spirals. *Proceedings 12th International Coal Preparation Conference*, Intertec Inc., Lexington, KY. 119–143.

MacNamara, L., Toney, T.A., Moorhead, R.G., Davies, P., Miles, N.J., Bethell, P., and Everitt, B. 1996. On site testing of the compound spiral. *Proceedings 12th International Coal Preparation Conference*, Intertec Inc., Lexington, KY. 253–276.

Mankosa, M.J., Stanley, F.L., and Honaker, R.Q. 1995. Combining hydraulic classification and spiral concentration for improved efficiency in fine coal recovery circuits. In *High Efficiency Coal Preparation*. Edited by Kawatra, S.K. SME, Littleton, CO. 99–107.

Mankosa, M.J. 1999. Personal communications. Eriez Magnetics. Eriez, PA, USA.

Mohanty, M.K., Honaker, R.Q., and Ho, K. 1998. Coal flotation washability: Development of an advanced procedure. *Coal Preparation*. 19: 51–67.

Nicol, S.K., Drummond, R.B. 1997. A review of experience with teeter bed separators. Preprints, Technology Review Seminar, Recent Developments in Gravity Separators, Australian Coal Preparation Society. May 21.

Osborne, D.G. 1988. Coal Preparation Technology. Graham and Trotman, Norwell, MA.

Reed, S., Roger, R., Honaker, R.Q., and Mankosa, M.J. 1995. In-plant testing of the Floatex density separator for fine coal cleaning. *Proceedings 12th International Coal Preparation Conference.* Intertec Inc., Lexington, KY. 163–174.

Richards, R.G., Hinter, J.L., and Holland-Batt, A.B. 1985. Spiral concentrators for fine coal treatment. *Coal Preparation.* 1: 207–229.

Riley, D.M., Firth, B.A., and Lockhart, N.C. 1995. Enhanced gravity separation. In *Proceedings of the high efficiency coal preparation: An international symposium.* Edited by Kawatra, S.K. Littleton, CO: Society for Mining, Metallurgy and Exploration, Inc. 79–88.

Salama, A.I.A. 1986. Yield maximization in coal blending. *Proceedings 10th International Coal Preparation Congress.* CIM, Edmonton. 196–216.

Snoby, R.J., Grotjohann, P., and Jungmann, A. 1999. Allflux—New Technology for Separation of Coal Slurry in the Size Range 3 to 0.15 mm. *Proceedings 16th International Coal Preparation Conference.* Intertec Inc., Lexington, KY. 126–139.

Venkatraman, P., Luttrell, G.H., and Yoon, R.H. 1995. Fine coal cleaning using the multi-gravity separator. In *Proceedings of the High Efficiency Coal Preparation: An International Symposium.* Edited by Kawatra, S.K. Littleton, CO, Society for Mining, Metallurgy and Exploration, Inc. 109–117.

Weldon, W.S., MacHunter, R.M.G. 1998. *Recent advances in coal spiral development.* Gold Coast, Australia: MD-Mineral Technologies, Inc.

SECTION 3

Non-coal Gravity Separations

- Heavy Media Separation (HMS) Revisited   **143**
- Recovery of Gold Carriers at the Granny Smith Mine Using Kelsey Jigs J1800   **155**
- Applications of the HydroFloat Air-assisted Gravity Separator   **165**
- Advances in the Application of Spiral Concentrators for Production of Glass Sand   **179**

Heavy Media Separation (HMS) Revisited

Roshan B. Bhappu[*] and John D. Hightower[*]

Although Heavy Media Separation (HMS) has been known for some time, the application of this versatile technology has not been appreciated by the mining industry up to now. With increasing capital and operating costs as well as environmental issues, it is quite apparent that the HMS concept whereas indicated should offer considerable benefits to the industry in the future. Preconcentration of metallic and industrial minerals at a relatively coarse crush size, prior to costly and energy consuming grinding operation, should result in discarding 30 to 70 weight percent of run-of-mine ore into the reject fraction. Thus, limiting the feed to costlier grinding and subsequent flotation or leach operation would be limited to a small fraction of the total plant feed. Such a flowsheet incorporating the preconcentration concept should result in improving the economics of the mining project substantially. This contribution provides the potential application of HMS for metallic as well as non-metallic minerals in light of current technological, economic and environmental issues.

INTRODUCTION

The Heavy Media Separation (HMS) process also referred to as the Dense Media Separation (DMS) process, in general, has been recognized as one of the most efficient gravity concentration methods to preconcentrate a large variety of minerals at a relatively low capital and operating cost. The preconcentrated product is then processed further by more expensive processes to recover the final product. With today's increasing labor, utilities and environmental costs for minerals handling and processing, it is imperative to discard the gangue components of the ore as early in the beneficiation process as possible, especially prior to fine grinding. In this endeavor, HMS technology certainly provides a potentially attractive preconcentration step in several flowsheets involving selective separation of metallic and non-metallic ores.

---

[*] Mountain States R&D International, Inc., Vail, Ariz.

Although known since the middle of last century and mostly applied to coal washing, HMS has not attained its true potential as a versatile step, prior to fine grinding, in conventional mineral processing as well as hydrometallurgical or pyrometallurgical flowsheets. This is primarily due to the lack of basic understanding of the process and the general notion that fine grinding of the total ore is the only method for achieving high recoveries, albeit at what cost. An in-depth scrutiny of this versatile HMS technology clearly reveals that this process can be effectively used for selective separation of diamonds and other gemstones; treatment of metallic ores of copper, lead and zinc; in recovery of gold and PGM minerals in some cases; and for separation of non-metallic and industrial minerals such as barite, fluorite, chromite, etc. Some more modern applications in non-mining operations include automobile scrap treatment, reusable resource recycling business, treatment of industrial and municipal wastes and waste from electronic industry.

In this paper, attempts will be made to emphasize the potential applications of this versatile HMS concept to the preconcentration of industrial minerals other than coal.

ADVANTAGES OF HMS

The most relevant advantages of the HMS concept are:

- Lower capital and operating costs (no grinding involved).
- Potential treatment of lower grade (below cut-off) resources; thus, increasing the overall ore reserves.
- Increase mine production without expansion of existing milling facilities.
- Allow non-selective mining at a much-reduced mining cost.
- HMS plant may be installed underground and the rejects (float product) may be used for backfill.
- Environmentally more friendly; disposal of coarse rejects on dumps, as backfill or as trap rock.
- In some cases, rejects may be used as by-products.
- Allow recovery of gemstones without hand picking with associated benefits in developing countries.
- Discard softer gangue minerals prior to grinding and causing deleterious slime problems in subsequent processing.

HMS TECHNOLOGY

HMS technology is based on the principle of separating valuable minerals from the gangue on the basis of their densities (sp.gr.). Under ideal conditions, separations may be achieved between two minerals varying in 0.2 sp.gr. differential. The selective separations occur not only with cleanly liberated mineral particles but also with middlings, thus, allowing high overall recoveries (+90%). Table 1 shows the densities of various marketable industrial minerals compared to typical gangue minerals. As can be seen, in general the sp.gr. of valuable minerals are at least 0.3 and higher, thus, assuring effective separation of the valuable minerals by HMS concept.

It is true that the higher sp.gr. minerals can be effectively separated by other gravity concentration methods such as tabling, spiraling, etc. However, these methods are applicable to finer sizes of feed below 28 to 35 mesh. What is unique about the HMS concept

TABLE 1 Densities of major industrial minerals and gangue

Mineral	Formula	Sp Gr.	Mineral	Formula	Sp Gr.
Barite	$BaSO_4$		Magnetite	$FeO \cdot Fe_2O_3$	5.2
Borax	$Na_2B_4O_7 10H_2O$	1.7	Serucite	MnO_2	4.8
Cassiterite	SnO_2	6.8–7.1	Rhodochrosite	$MnCO_3$	3.8
Celestite	$SrSO_4$	3.9–4.0	Rutile	TiO_2	4.2
Chromite	$FeO \cdot Cr_2O_3$	4.3–4.6	Scheelite	$CaWO_4$	5.9–6.1
Columbite	$(FeMn)(CaTa)_2O$	6.3	Spodumene	$L:Al(SiO_3)_2$	3.2
Corundum	Al_2O_3	3.9–4.1	Strontianite	$SrCO_3$	3.7
Ferberite	$FeWO_4$	7.2–7.5	Sulfur	S	2.0
Fluorite	CaF_2	3.0–3.3	Tentalite	$FeTaO_6$	5.3–7.3
Gypsum	$CaSO_4 \cdot 2H_2O$	2.3	Uraninite	Variable	9.0–9.7
Hematite	Fe_2O_3	4.9–5.3	Vanadite	$(PbCl)Pb_4(VO_4)_3$	6.6–7.1
Huebnerite	$MnWO_4$	7.2–7.5	Wolfranite	$(FeMn)WO_4$	7.2–7.5
Ilmentite	$FeTiO_3$	4.5–5.0	Zircon	$ZrSiO_4$	4.2–4.7
Magnesite	$MgCO_3$	3.1			
Major Gangue Minerals			**Major Gem Minerals**		
Quartz		2.65	Diamond		3.2–3.5
Feldspars		2.6–2.7	Emerald		2.75
Calcite		2.75	Ruby (Spinel)		3.5–4.1
Dolomite		2.8–2.9	Topaz		3.4–3.6
Clays		2.6	Aquamarine		
Serpentine		2.5–2.6	Garnet		3.2–4.3
Talc		2.7–2.8			
Fe:Silicates		2.8–3.0			

is that such effective separations are achievable even at relatively coarse particle sizes up to 2-½ inch for cycloning devices to as coarse as 6 to 8 inch for static drum type separations. Thus, the HMS step can be incorporated into the flowsheet either on the primary, secondary or in some cases tertiary crushed products. Also, in industrial practice, the feed to the HMS circuits are screened to remove the minus 28- or 35-mesh products not because of process ineffectiveness but in order to reduce the loss of relatively expensive heavy media (Fe:Si) used in the process.

LABORATORY TESTING

In designing a workable flowsheet incorporating the potential HMS step, it is essential to conduct a realistic laboratory sink-float testing program to determine the optimum separation size, density and other operating parameters on a representative sample. These tests are performed by immersing the predetermined crushed sample in a series of heavy liquids at varying densities and analyzing the sink and float products to obtain a metallurgical balance. The tests are then repeated at coarser or finer sizes in order to obtain the best results. Once the optimum parameters are ascertained, a confirmatory sink-float test on a larger sample is conducted using these parameters and a conceptual flowsheet with material balance is developed.

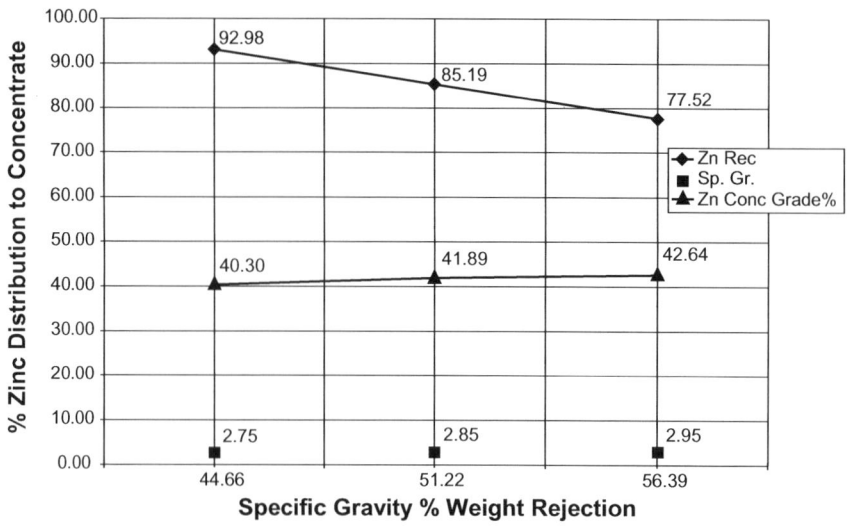

FIGURE 1 Mexican zinc-oxide composite sample –1" crush feed grade 24% Zn. Zn distribution (%) vs. weight % rejection with varying liquid densities

It is strongly recommended that the preliminary sink-float test program be continued into a pilot plant campaign in order to determine the operating and design parameters for the process. Since these pilot plant tests are usually conducted with the heavy media rather than the heavy liquid, the results are more realistic and duplicate the commercial plant practice. The data collected from the pilot test can now be used for the final feasibility study and plant design and construction.

The heavy liquids used in the laboratory sink-float tests were predominantly halogenated hydrocarbon such as acetylene tetrabromide (3.0 sp.gr.). Because of the carcinogenic nature of these liquids, the current trend is to use environmentally friendly inorganic heavy liquids using cesium and rubidium compounds. It is also possible to obtain higher density liquids by suspending metal powders in heavy liquids (up to 3.6 to 3.7 sp.gr.).

The results of a typical sink-float test program are shown in Figure 1.

PILOT PLANT TESTING

Based on these encouraging results of the sink-float test program, a more detailed pilot plant testing campaign is undertaken using the Dense Media Separation (DMS) units using Fe:Si, magnetite or their mixtures depending on the desired media density. The HMS pilot plant units include Heavy Density Cone Separators, Heavy Media Cyclones and the Dyna Whirlpool (DWP). The capacities of these pilot plant units range from a few hundred pounds to as much as 3 tphr for a 6-inch DWP unit. As stated earlier, the results of the pilot plant testing are used for engineering design and commercial operation.

In general, the pilot plant tests are conducted on the best crushed size and plus 28, 35 or 48 mesh at the media density indicated by sink-float tests. Additional tests are run at lower as well as higher than the indicated density, since the use of Fe:Si under a centrifugal force provides somewhat different separation conditions as compared to heavy liquids. Other dense media such as magnetite (sp.gr. of 5.18), galena (sp.gr. of 7.8), barite (sp.gr. of 4.5), etc., can be used. However, Fe:Si with or without magnetite is the

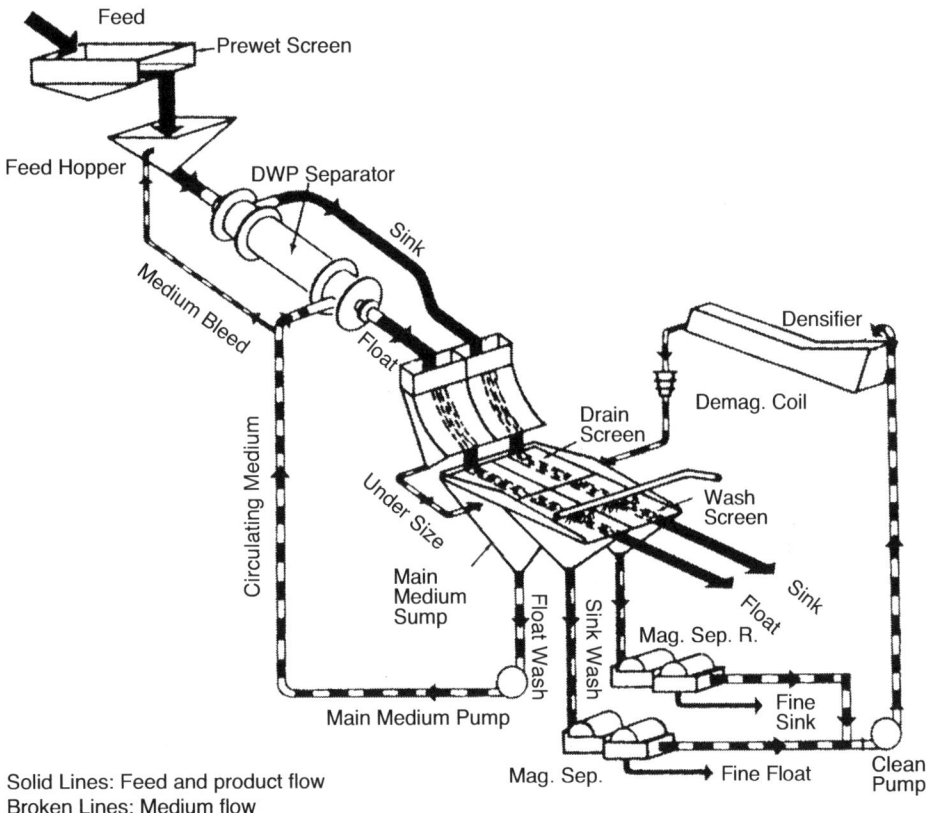

FIGURE 2 Typical HMS flowsheet

preferred media due to ease recovery of media by magnetic separation step in the process. The loss of heavy media in HMS process is about 1 to 1.5 lbs/ton feed.

EQUIPMENT

The basic flowsheet for HMS process is shown in Figure 2. The major steps in the process are:
- Feed preparation;
- Separation;
- Separation of media from the products; and
- Reclamation and recycling of the media.

As stated previously, the ore is crushed to the optimum size indicated from the initial sink-float tests and pilot plant data. The HMS plant feed is usually wet screened to remove the minus 35-mesh product which is diverted to the subsequent grinding and processing by froth flotation, agitation leach or other appropriate process. About 5 gpm of water per ton is used for wet screening.

The separatory vessel is the most crucial part of the HMS system and may consist of a drum, a conical tank or a cylindrically shaped vessel typified by dense media cyclone

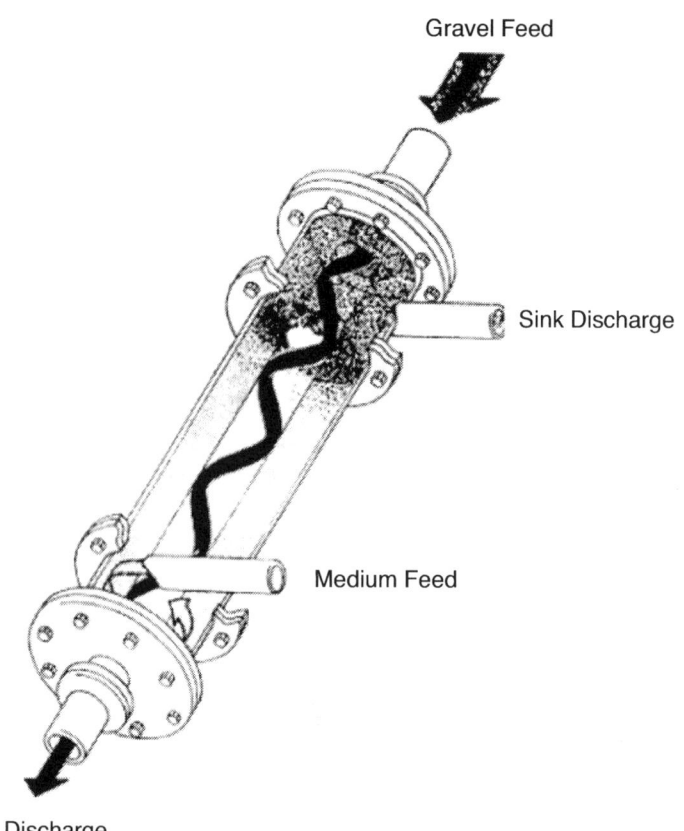

FIGURE 3 Dyna whirlpool material flow

or DWP. In this latter case (Figure 3), the process is based on a method of creating and controlling cyclonic motion, utilizing the separating characteristics of the vortex. This versatile, rugged and compact system is operated in conjunction with other standard auxiliary equipment (screens, magnetic separators, etc.) to allow a low-cost, space-saving plant construction. It requires minimum operational supervision and is suitable for complete automation resulting in very low operating costs.

Once the sink-float separation is achieved in practice, these products are then screened and washed separately to remove the heavy media (Fe:Si). The heavy media in the screened undersize is then recovered by magnetic separation. The magnetic product is next desified in a spiral classifier and then returned to the stock media tank where the density of the media is regulated to the desired level. Finally, the pre-adjusted media is recycled to the process. As can be seen, the entire HMS process is relatively simple, workable and very cost effective.

COMMERCIAL APPLICATIONS

Because of the effectiveness of separation of two minerals varying in sp.gr. of 0.2 and above, the HMS technology is widely applicable for separating several industrial minerals in practice with associated reduction in capital, operating and environmental costs.

As can be seen from Table 1, the HMS technique may be utilized not only for processing many industrial minerals, such as barite (4.5) and chromite (4.4) from the typical gangue, quartz (2.65) and calcite (2.7) where the sp.gr. differential is more than 0.5 but also when the differential is only 0.3 to 0.4 sp.gr. such as separation of fluorite (3.2) and magnesite (3.1) from quartz and calcite. Moreover, the HMS concept also provides an effective process for separating valuable industrial minerals lower in sp.gr. than the gangue minerals such as potash (2.0–2.1) from shale, limestone and sandstone (2.6 to 2.70). In this instance, the float product is the valuable component. Other applications include separation of halloysite (2.0–2.2), gibbsite (2.4), beryl (2.6–2.8), borax (1.7), brucite (2.4), carnalite (1.6), chyrsotile (2.2), gypsum (2.3), kainite (2.1) and sulfur (2.0).

In recent years, the application of HMS technology has been widened to include effective separation of bauxite (2.6) from lateritic iron (3.6–3.8) and separation of oxide minerals of lead, zinc and even copper. In one study, Roshan Bhappu et al. have shown that the New Mexican pegmatite deposit containing spedumene (3.1–3.2), muscovite (2.9) and beryl (2.8) can be selectively separated from quartz (2.6) and feldspars (2.6–2.7) under appropriate conditions using multiple stage HMS processing. Another unused example of the use of HMS concept is the removal of troublesome gangue components (talc, serpentine, etc.) from PGM ores as well as preconcentration of PGM from submarginal and mine waste. In addition, in some cases the disseminated gold values from their ores have been preconcentrated by selectively separating the host rocks (quartz, calcite or iron oxides) from the associated gangue rock.

Finally, more recently it has been conclusively shown by Mountain States R&D International, Inc. (MSRDI) that just as HMS application to diamonds (3.2–3.5), this versatile technique can be used for recovery of precious stones such as emeralds (2.75), rubies (3.5–4.1), topaz (3.4–3.6), zircon (4.2–4.7) and garnet (3.2–4.3). Such innovative applications of HMS technology to precious stones and gems has resulted in increasing the mine production from a few tons by manual labor to as much as several hundred tons per day with partial mechanization resulting in improved economics and reduce pilferage.

EXAMPLES FROM PRACTICE

The partial list of commercial applications of the HMS concept for processing industrial minerals over the last 25 years is listed in Table 2. As can be seen, a wide variety of non-metallic ores are treated effectively to obtain preconcentrations of the valuable minerals by discarding 20 to 60 weight percent of the gangue float fraction at a relatively coarse crushed size. The recoveries in the peconcentrated products (sink fraction) are relatively high with minimum loss of values (5 to 15%) in the reject. Such a technique results in appreciable reduction in capital and operating costs along with attractive environmental benefits related to tailings disposal, less dust loss, lower utilities costs, etc. Moreover, in some cases, the HMS plants have been located underground with the disposal of float (waste) as a backfill.

The application of HMS technology to processing of diamondiferous ores taking advantage of the higher sp.gr. of diamonds (3.2–3.5) from the alluvial or kimberlite gangue is now a standard commercial operation worldwide, and many more new operations are scheduled to go into production in the near future. Though, the application of

TABLE 2 Partial list of HMS processing plants

Operator/Plant Location	Mineral Processed	Size Range	Plant Feed tph	HMS Units	Sink/Float Ratio	Separa. Density
Aluminum Co. of Canada, Ltd. St. Lawrence, Newfoundland	Fluorspar Barites	–3/4"+20M	80	2–15"	40/60	2.72
Barton Mines North Creek, NY	Garnet	–1/4"+45M	60	1–12" & 1–9"	60/40	3.2
Basic Inc. Gabbs, NV	Magnesite	–3/8"+20M	30	1–9"	75/25	2.9
Bethlehem Steel Corp. Icomi Mine, Amapa, Brazil	Manganese	–1/4"+20M	130	2–15"	80/20	2.9
Cia. Minera de Autlan S.A. Autlan Mine	Manganese	–20mm +1mm	90	1–18"	70/30	2.95
Universe Tankships Inc. Para, Brazil (Jari, Project)	Bauxite	–3/8"	18	1–9"	95/5	2.3
Companhia Mineira do Lobito Jamba Mine, Angola, Africa	Iron	–1/4"+20M	400	6–15"	80/20	2.7
Dresser Minerals Ryder Point Plant	Fluorspar Barites	–25mm +1mm	80	2–15"	50/50	2.75
Fundy Gypsum Co. Ltd. Windsor, Nova Scotia	Gypsum Rock	–3/4"+20M	80	2–15"	30/70	2.5
International Mining Co. Enramada Mine, Bolivia	Tin/ Tungsten	–1"+20M	30	1–12"	25/75	2.95
Lithium Corp. of America Bessemer City, NC	Lithium	–1/4"+65M	65	1–15"	30/70	2.8
NL Industries Inc. Guatemala	Scheelite	–1/2"+14M	10	1–9"	60/40	2.7
Renison Ltd. Zeehan, Tasmania	Tin	–1/2"+28M	80	2–15"	80/20	3.0
Southern Peru Copper Corp. Ilo, Peru	Coquina Shells	–1/4"+30M	50	1–15"	50/50	2.7
Turk Maadin Sirketi Beyoglu, Turkey	Chrome	–1/4"+20M	10	1–9"	60/40	2.9

HMS concept for diamonds is well understood and documented, the application of this versatile technique for concentration of other gemstones such as emeralds and rubies have not been investigated or reported. Recently, Bateman Engineering has reported the availability of 1 to 5 ton/hr dense media separation (DMS) modular unites for treating diamond and emerald ores in South Africa. Several years ago MSRDI, supported by UNDP funds, had developed a unique flowsheet involving HMS for treating emerald bearing ores in Northern Pakistan and Afghanistan. The finalized flowsheet for this application is shown in Figure 4. Using this HMS concept it was possible to treat emerald ores containing about 5 to 6 carats per ton of gemstones.

ECONOMIC CONSIDERATION

The HMS process, because of the simplicity of operation and equipment, constitutes one of the cheapest methods of mineral concentration. The primary reason for the cost advantage is that the specified preconcentration is carried out at a relatively coarse size

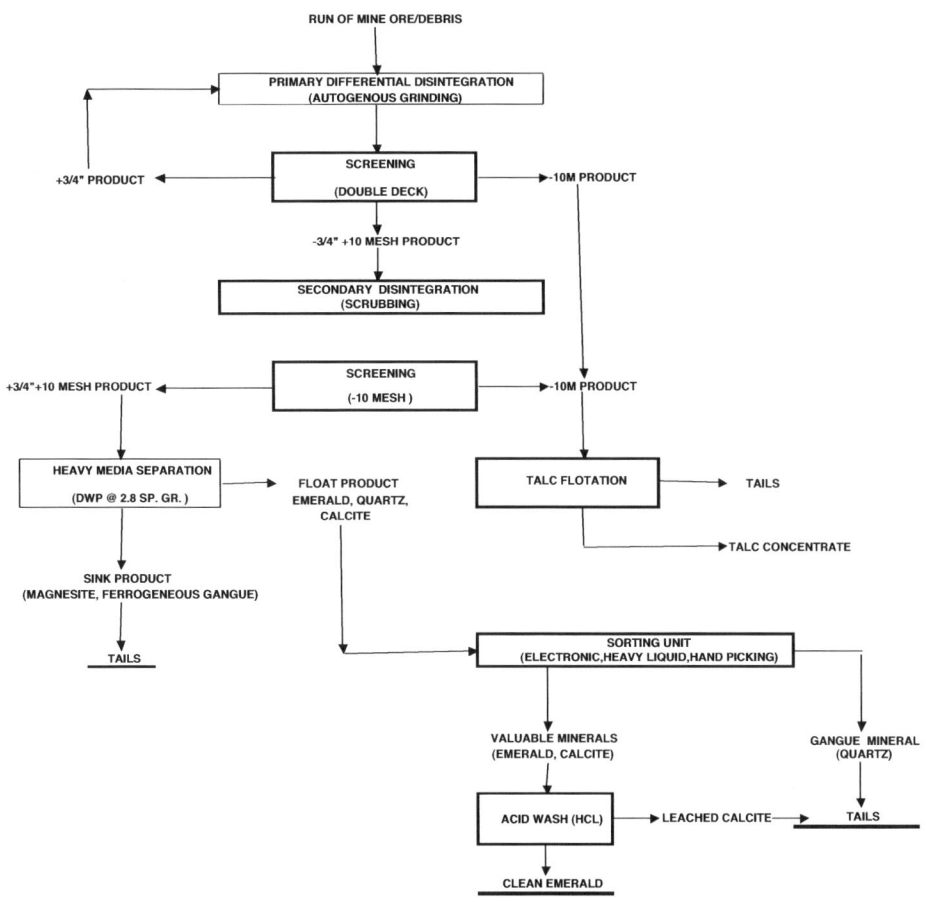

FIGURE 4 Conceptual flowsheet for emerald recovery

prior to fine grinding which is the most expensive step in the milling operation. Moreover, the HMS separation results in discarding substantial portions of the coarser plant feed, which reduces the cost of the tailings disposal and has other environmentally attractive features. In the case of underground mines, the plant may be installed underground and the float discard can be used as a backfill.

The installed capital cost of HMS plants is dependent on several factors including plant size, feed preparation requirements, the size range of the feed and auxiliary facilities such as plant building utilities availability and hook-up and the like. In general, the larger the plant capacity the lower the installed cost per ton of ore processed. In practice, the HMS units are rated at tons per hour of ore processed and number of units are determined based on the size of the HMS cyclone and the total feed per day. In the case of the DWP Separator, Table 3, the size of the unit and the corresponding tonnage per hour as well as tons per day at 90 percent availability.

An added advantage of the HMS system is the relatively small plant area requirement with respect to the tonnage processed. A 30 tph plant (650 tpd) will cover about 800 square feet while a 125 tph plant (2,700 tpd) will cover about 1,500 square feet. The equipment is usually located on two to three floors, including the basement and is

TABLE 3 HMS (DWP) plant capacities by size (90% availability)

Cyclone Size (Inch-Diameter)	Application	Capacities	
		STPH	STPD
6	Pilot Plant	3–5	65–108
9	Commercial	15	324
12.5	Commercial	35	756
15	Commercial	62	1340
18.5	Commercial	100–110	2,160–2,375
4–18.5" Units	Commercial	400–440	8,840–9,500

arranged to maximize gravity flow of the ore and the media. As pointed out earlier, the HMS plant may be located underground with associated benefits.

The operating cost of HMS operations are also economically very attractive and include cost of labor, utilities, supplies and the media losses. Since the HMS operations are relatively simple they can be readily automated with one man operating several parallel circuits. Thus, the most significant cost savings, especially for higher tonnage plants, is the cost of labor per ton of feed.

In general, the costs such as power, water, maintenance supplies and media loss are more or less linear than the labor cost with respect to unit costs based on feed rates. Most of the power requirement (60 to 70%) is consumed by pumping of the media while the remainder will be consumed by conveying, magnetic separation, screening, etc. Maintenance costs average about 20 to 25 percent of the total operating cost. On the other hand, feed size and the abrasiveness of the products are major factors effecting maintenance costs. Make up water requirements are very reasonable with about four gallons per minute per ton of feed.

Moreover, it should be noted that the cost of media losses is one of the major cost items in the operating cost. The loss of media amounts to about 0.5 to 1.5 lbs per ton of feed. At the relatively high cost of Fe:Si ($600 per ton), the media loss could be as high as $0.45 per ton of feed. Thus, out of the total operating cost of about $1.50 to $3.00 per ton, the cost of media losses may amount to 25 to 45 percent of the total cost. Accordingly, every effort must be made to reduce the media losses through proper design and appropriate magnetic separation devices.

The capital and operating costs for the HMS operations are provided in Table 4. The overall installed capital cost includes the required crushing and screening plant to prepare the HMS plant feed. In the event the HMS plant, instead of a stand-alone facility, is an integral step in the flowsheet leading to fine grinding and flotation, the cost of the HMS operation as a fraction of the overall cost would be reduced significantly. The table also includes the comparative cost of processing with and without the inclusion of the versatile HMS circuit in the flowsheet. As can be seen, the resultant savings in both the capital and operating costs with the HMS step are significant. As can be appreciated, these cost savings would alter the overall economics of the project from a mediocre to a very viable venture or from a marginal project to an economically viable one.

TABLE 4 Estimated capital and operating costs of HMS plants

Capacity (STPD)	Capital Costs		Operating Costs (s/t)	
	Without Crushing	With Crushing	Without Crushing	With Crushing
500	$1,000,000	$3,000,000	$2.00	$3.20
1,000	$1,300,000	$4,000,000	$1.50	$2.40
5,000	$2,500,000	$7,000,000	$1.20	$2.00
10,000	$3,750,000	$10,500,000	$1.00	$1.50

Comparative Costs Fluorspar Operation—5,000 STPD @ 33% CaF2

Basic Flotation Plant		HMS + Flotation Plant	
Capital Cost	Operating Cost	Capital Cost	Operating Cost
$35,000,000	$10.00/T	$20,000,000	$6.00/T

CONCLUSIONS

Based on the above discussions concerned with the practical and economic viability of the inclusion of the versatile HMS technology for industrial mineral processing, it is concluded that this preconcentration concept should find increasing applications in the future. The major advantages include:

- Lower capital and operating costs, which could significantly influence the overall economic viability of the project.
- Feasibility of processing lower grade ores thereby increasing the overall old reserves, mine life and ultimate conservation of mineral resource.
- Potential installation of the HMS plant underground with associated benefits including use of rejects as backfill.
- Environmental advantages concerned with tailings disposal, water and power conservation and in general, an environmentally more attractive processing option.

It is hoped that the industrial minerals operations would take a closer look at this versatile industrial mineral processing option for their existing operations as well as other grassroots projects.

Recovery of Gold Carriers at the Granny Smith Mine Using Kelsey Jigs J1800

G. Butcher[*] and A.R. Laplante[†]

The Granny Smith Gold Mine is faced with a new and more refractory ore source in its Wallaby deposit. The increasing presence of finely occluded gold in pyrite necessitated an upgrade of the existing gravity circuit comprising Reichert cones and spirals. Evaluation of the current generation of continuous, high throughput centrifugal concentrators led to the selection of the Kelsey centrifugal jig for the duty. This paper tracks the development of the circuit from laboratory, pilot and full scale field evaluation to the commissioning in mid-2002 of the current installation of three Kelsey jigs treating some 250–300 tph of CIP plant tailing.

INTRODUCTION

The Granny Smith Gold Mine, located in Western Australia, is managed by Placer Dome, who has 60% equity in the operation. Gold intimately associated with sulphide minerals is typically concentrated using froth flotation. This unit process, however, adds an often unacceptable level of complexity to the conventional cyanide leach-CIP circuit, especially when these sulphides are finely disseminated and the mill feed comprises oxide: fresh ore blends containing clay minerals as experienced at Granny Smith. Flotation reagents can also interfere with carbon activity, which is why gravity circuits are generally chosen to recover gold carriers from cyanidation residues for regrind and additional cyanidation, typically at increased cyanide concentration. The recovery of semi-refractory gold associated with predominantly pyrite had previously been achieved at Granny Smith through the use of conventional cone and spiral concentrators, which is typical of tailings retreatment plants in Australia (Martins, Dunne and Delahey, 1993). A high-grade heavy mineral fraction was recovered from deslimed CIP tailing for reprocessing by fine grinding

[*] Placer Dome Asia Pacific, Brisbane, Queensland, Australia.
[†] Dept. of Mining, Metals and Materials Engineering, Montreal, Quebec, Canada.

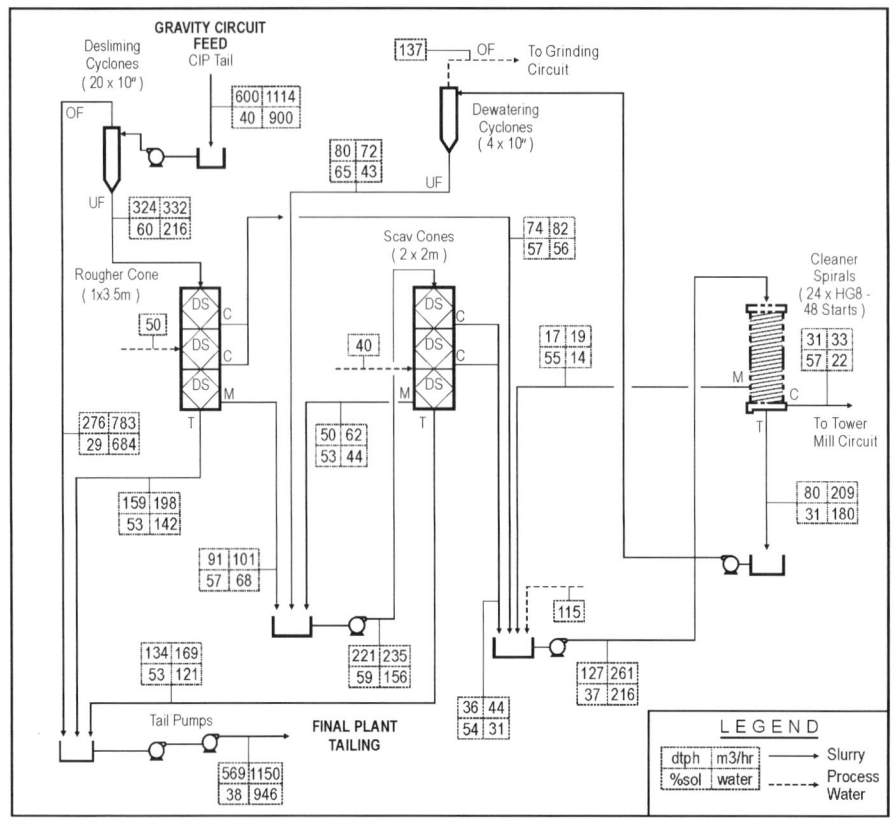

FIGURE 1 Flowsheet and typical mass balance of original tailing retreatment circuit

and re-leach. The recovered gold typically comprised some 2–3 percent of that contained in plant feed representing an increment of approximately 10,000 oz gold per annum. Figure 1 shows the flowsheet of the original circuit and presents a typical mass balance.

Gypsum and carbonate scale build-up on the existing Reichert cones, as a consequence of adding lime to saline process water (total dissolved solids between 15,000 g/t and 60,000 g/t), resulted in poor feed distribution with consequent deterioration in metallurgical performance and excessive maintenance costs. In 1998 the nearby Wallaby ore deposit was discovered and metallurgical evaluation commenced on what was to become a 58Mt resource grading 2.65gAu/t. Conventional heavy liquid separation of cyanide leach residue at a grind size P_{80} of 140 μm during scoping test work confirmed that the new deposit had a greater gold-pyrite association with finer grained sulphide minerals than previous ores. This necessitated a significant upgrade of the gravity circuit to achieve optimum gold recoveries, and a testing programme was commenced to determine the plant requirements. This contribution describes the test work that led to the decision to install a Kelsey jig-based gravity circuit.

INITIAL TEST WORK

The standard hydrometallurgical test protocol employed on Wallaby ore is shown in Figure 2 with the gravity pre-concentration step primarily designed to eliminate the problems associated with the nugget effect of coarse free gold in the leach feed. Typically

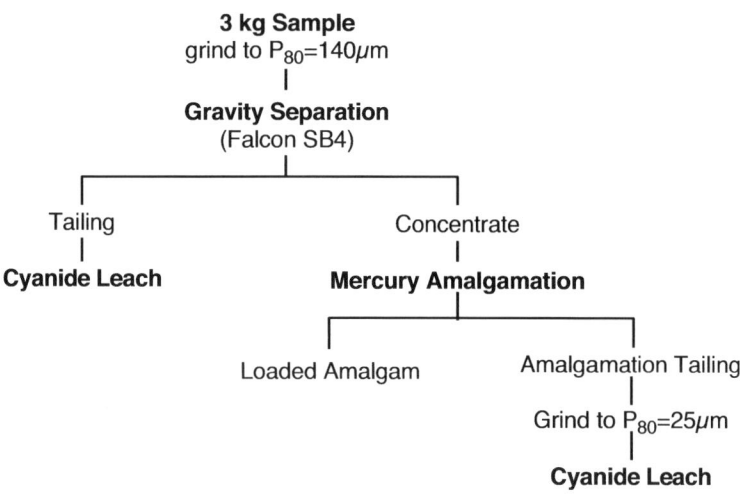

FIGURE 2 Test protocol

TABLE 1 Gold distribution in leach residues

	Metal Distribution (%)								
	4.05 < S.G.		3.30 < S.G. < 4.05		2.85 < S.G. < 3.30		S.G. < 2.85		Slimes (–45 μm)
Ore type	Au	S	Au	S	Au	S	Au	S	Au	S
Fresh	49	37	5	5	8	9	15	10	23	39
Oxide	11	30	19	7	17	5	32	18	21	40

TABLE 2 Residue concentrate leach results

				Au Extraction @ Hours (%)			
Ore Type	Mass Yield to Conc. %	Calc'd Head (gAu/t)	Grind Size (μm)	2	4	8	24
Fresh	3.2	9.3	25	69	75	79	80
Oxide	5.9	0.8	25	69	72	74	79

thirty-five percent of the total gold in feed was recovered from the amalgamation tailing containing some sixty percent of the sulphur in the feed.

At a grind size (P_{80}) of 150 μm, scoping test work showed that the Wallaby sulphide ores yielded cyanidation recoveries ranging from 70% to 80%, much lower than the oxide ore recovery of 96%. Table 1 shows how gold and sulphur are distributed in the cyanidation residues of both the oxide and sulphide components of the Wallaby ore. The weight of the +3.30 SG fraction is typically less than 3.0% and contains a significant portion of the residual gold (54% for fresh ore and 30% for oxide ore).

Bulk leach residues samples from five ore type composites were subjected to dense liquid separation at 2.96 SG. The concentrates were then ground to a P_{80} of 25 μm and cyanided. The leach results for all composites indicated rapid leaching with gold recoveries around 80%. Averaged results are shown in Table 2.

TABLE 3 Statistical representations of 2000 data, third test series

	Two-product Formula		Balanced Data		Regression
	Average	Std. Error	Average	Std. Error	Std. Error
Sulphur Recovery, %	52.6	9.6	46.3	6.9	2.5
Gold Recovery, %	43.0	24.7	39.8	8.4	4.8

The test work indicated that Wallaby fresh ore was amenable to increased gold extraction by concentrating by gravity a low mass, gold-rich stream from leach tailing and subjecting this concentrate to fine grinding and cyanidation. The lower recoveries of primary cyanidation at a P_{80} of 140 µm for sulphide ore types provided the economic incentive to improve the existing Reichert-cone based circuit. In particular, gold carriers between 25 and 75 µm were now targeted, making the use of centrifuge units very desirable, if not essential.

PILOTING THE PROPOSED GRAVITY CIRCUIT

Kelsey jigs, in-line pressure jigs and Falcon C-series concentrators had been evaluated by Granny Smith as possible replacement units for the Reichert cones, prior to the Wallaby discovery and initial test work. In-line pressure jig results were unacceptable, whereas both Falcon C400 and Kelsey J200 results were encouraging, but could not justify the capital cost of a new circuit with the existing mill feed. The advent of the Wallaby project provided the economic incentive for the circuit upgrade. On the basis of lower capital costs and the availability of a 1.80 metre diameter J1800 model for evaluation at moderate cost, the Kelsey jig was selected for extensive on-site evaluation during 2000 and 2001. The J1800 was initially rated for 40 to 60 t/h, and the 2000 on-site test work was aimed at measuring the performance of the J1800 in this feed rate range and evaluate the potential for achieving acceptable recoveries at a higher feed rate, to lower the relatively high capital cost per tph. The first two test series were aimed at tuning the unit, in particular feed distribution. A third test series, with an improved feeding system, yielded sixteen valid tests. Recoveries obtained using the two-product formula were very noisy (Table 3), on account of the low recoveries achieved, typical of scavenger circuits (Laplante, 1984). No clear link between recovery and yield was apparent. Mass balancing using the mass flow data as well as sulphur and gold analyses cut down recovery variance significantly, as shown in Table 3.

Further linking the recovery to yield and feed rate yielded the following regressions (only the parameters with a better than 95% significance were retained):

$$S \text{ Rec. in \%} = 11.1 + 0.46 * \text{pulp feed rate} - 0.67 * \text{dry feed rate} + 3.8 * \text{yield}$$

$$Au \text{ Rec. in \%} = 0.24 * \text{dry feed rate (t/h)} + 3.5 * \text{yield}$$

Both regressions have significance in excess of 99.9%. Their standard error is significantly lower than that of the balanced data (Table 3). They indicate that within the range of feed rates tested, 35 to 55 dt/h, increasing feed rate has no detectable deleterious effect on recovery. Further, there is now a clear link (as expected) between yield and recovery, hence the logical next step, additional testing at higher feed rate and higher yield. This would require increasing of the jig screen apertures and the ragging size, and evaluating two-stage concentration, since the jig concentrate rate would be too high for the existing vertimill capacity. The optimisation process led to two pilot campaigns in January and February 2001, with a bleed stream of Reichert cone feed diverted to the test unit fitted with 1000-μm aperture screens. While the operating characteristics and mechanical reliability of the device was being evaluated on this unit, parallel studies were being undertaken at a laboratory and pilot plant level on material from the Wallaby deposit. These studies confirmed that a rough concentrate obtained with a Kelsey J200 could be upgraded with over high recoveries using a Mozley laboratory separator.

The January test series averaged a dry feed rate of 80 t/h, whereas the February series was performed at a slightly higher average feed rate, 100 t/h. Yields averaged 15% for January and 18% for February. Recovery scatter showed the same trends observed with the third series of the 2000 data, as the standard error (standard deviation) decreased significantly with mass balanced data. Sulphur recovery tracked gold recovery well, but was on average 2.7% higher. All data (40 tests, 2 elements) could be represented with the following relationship:

Gold recovery (or sulphur recovery minus 2.7%) =

$53.3 - 0.327 \pm 0.060 *$ Feed Rate $+ 0.0146 \pm 0.0027 *$ Feed Rate * Yield

The regression has a very high significance (0.000011), with a mean residual (standard error) of 4.9%. It predicts that at feed rate of 80 t/h and a yield of 20%, gold recovery would be equal to 50%, and sulphur's to 53%. Processing a bulk concentrate sample of the J1800 jig with a J200 jig confirmed the excellent sulphur upgradability, with sulphur recoveries of 94 to 96% at yields of 25 to 45%. Gold recoveries were lower, 84 to 89%, but this was expected, as the early 2001 mill cyanidation residues had a much weaker association of gold with sulphides than what has been observed with the Wallaby sulphide ores during scoping work.

CIRCUIT DESIGN

Design Considerations

The Granny Smith process plant recovers in excess of 1,000 ounces of gold per day so it was important that disruption to existing operation was kept to a minimum. For this reason a separate facility was erected to house the Kelsey jigs and associated infrastructure. Figure 3 shows the jig circuit flowsheet.

Feed Rate. Wallaby fresh ore, which will comprise the bulk of the process plant feed, is exceptionally hard with UCS values as high as 350 Mpa. However, the initial feed supply from the upper, highly weathered levels of the deposit is very soft, so process plant milling rates were predicted to vary from 360 to 600 tph. Although the jigs have a nominal design feed rate of 85 tph per unit, this could increase to 100 tph on certain ore blends. The units had been shown during field trials to operate at feed rates up to 120 tph, but there was insufficient time to confidently establish metallurgical performance under those conditions. This was not considered critical, as the highest feed rates would be obtained with a high content of oxide ore, whose overall recovery is less dependent on the jig circuit.

Feed Sizing. The desliming cyclone split size was also an important consideration due to the potential to establish a significant circulating load of fine sulphides with the reground gravity concentrate being returned to the main leach circuit. The optimum

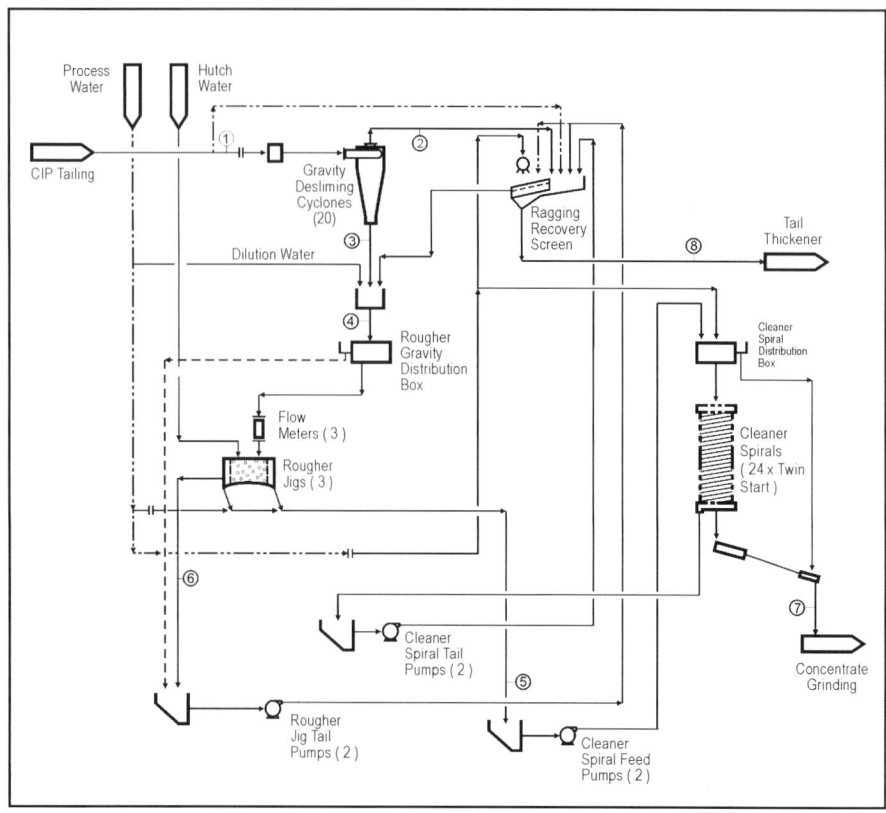

Stream Number	1	2	3	4	5	6	7	8
Stream Description	Desliming Cyclone Feed	Desliming Cyclone O/f To Tail	Desliming Cyclone U/F	Total Jig Feed	Jig Concentrate	Jig Tail	Spiral Concentrate	Gravity Tail To Thickener
Mass Flow - Solids (t/h)	385	145	240	240	48	192	17	368
- Liquids (t/h)	572	387	185	185	119	222	17	723
- Slurry (t/h)	957	532	425	425	167	414	34	1092

FIGURE 3 Flowsheet of the jig circuit

regrind size was indicated by the grind versus cyanide soluble gold response data shown in Figure 4. The ultimate regrind size, however, is dictated by the installed power of 375 kW in the existing vertimill.

The top size of the jig feed was also important in that blinding of the jig screens warrants special design attention due to the potential for high out-of-balance forces to be generated at the jig screen at rotation speeds of 170 rpm (28Gs at the screen). The wedge wire jig screen aperture of 1,000 μm installed at Granny Smith is much larger than that typically used, primarily to achieve the high concentrate mass yields required. The aperture of the trash and carbon safety screens was reduced from 1,000 to 800 μm to minimize jig screen blinding.

Similarly jig hutch water and jig screen cleaning system water needed to be free of particulates, a potential problem in arid regions where process waters are highly recycled. This necessitated installation of multistage in-line filters.

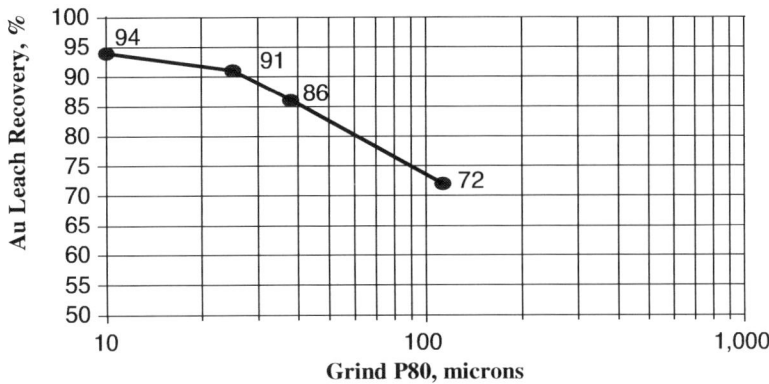

FIGURE 4 Sulphide concentrate grind optimisation

TABLE 4 Design criteria for gravity roughing circuit

Gravity Roughing				Low sulphide ore	Design
Equipment	Type		Kelsey J1800 jig		
	Number	Operating		3	3
	Solids capacity	t/h/unit		110	80
	Concentrate mass pull	%		20	20
	Hutch water*	m³/h/unit		51	51
	Total hutch water	m³/h		156	156
	Hutch water to conc.	% total		88	76
	Hutch water to tails	% total		12	24
Rougher Conc.	Solids	t/h		66	48
	Slurry	% solids		32	29
Rougher Tailing	Solids	t/h		264	192
	Slurry	% solids		49	46

* Hutch water flow rate in operation is 60 m³/h

Design Criteria

Circuit Description. The CIP tailing is deslimed to remove nominally minus 25 µm fines using cyclones ahead of the gravity circuit. The gravity circuit comprises a jig roughing stage and a spiral cleaning stage. The cleaner concentrate is ground in a vertimill operated in closed circuit to a nominal grind size P_{80} of 38 µm and then subjected to intensive cyanidation. The tailing from the intensive cyanidation is recirculated to the leach feed.

Centrifugal Jig Circuit. The design criteria for the Kelsey J1800 installation are shown in Table 4. "Low sulphide ore" represents a blend of fresh and oxidised ores in the mill feed.

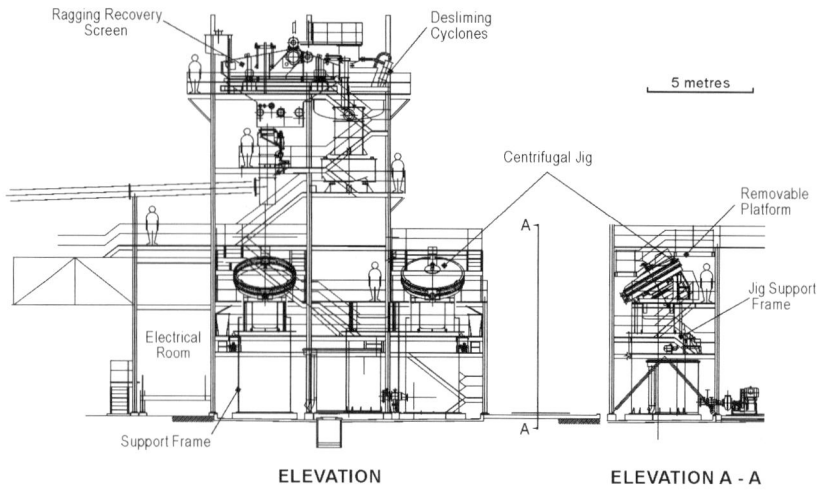

FIGURE 5 Jig circuit layout

Circuit Layout. The general arrangement of the jig plant in elevation is shown in Figure 5.

Both the desliming of the jig feed and the recovery of the ragging are performed above the three jigs, to gravity feed the jigs via a common feed well. Accessibility to the jigs is achieved via a removable platform.

DISCUSSION

Figure 6 illustrates why the jigs prove effective for the duty: sulphur and gold recovery are maximum between 30 and 200 µm. Above 200 µm, gold and gold-bearing sulphides are present in low concentrations and are poorly liberated; below 30 µm, recovery drops because the retention time in the unit is too low for effective separation, given the terminal velocity of gold and gold carriers, even in the centrifuge field generated by the J1800.

Successful application of the J1800 jig to this duty was by no means straightforward, as much thinking "outside the envelope" went into the test work. First, this is the first commercial application of the J1800 jig (although the unit had been extensively tested for iron ore scavenging in Eastern Canada). Second, this is also, to the knowledge of the authors, the first application of the centrifuge jig to a scavenging application; typically, the smaller capacity unit, the J1300, is used in beach sand (kyanite rejection from zircon) and for cassiterite and tantalite recovery (Beniuk, Vadeikis and Enraght-Moony, 1994). Earlier prototypes of the Kelsey had been tested for scavenging applications before (e.g., Wyslouzil, 1990), but none had yielded sustained commercial operation. Third, typical centrifuge jig applications are aimed at producing very high-grade or even final concentrates. To increase capacity at only a moderate loss of recovery, yields were increased, to the detriment of upgrading. The jig concentrate has since proven a very good candidate for upgrading in the existing spiral circuit at Granny Smith, validating this approach. This is largely due to the dynamic screening capability of the jig, which

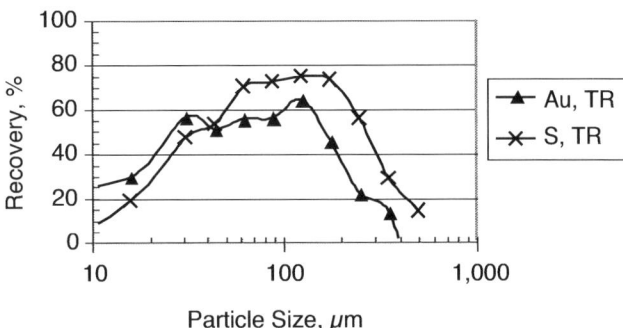

FIGURE 6 Size-by-size sulphur and gold recovery, Kelsey J1800 (from the 2000 third test series)

TABLE 5 Summary of test work program

	Existing Ore	Wallaby Ore
Dense liquid tests on leach residues, scoping and feasibility studies		See Tables 1 and 2
Cyanidation of reground heavy fractions		
Bench-scale recovery of gold carriers with J200 Jig	Prior to the feasibility study of Wallaby deposit	On bulk leaching residue sample
Piloting of J1800 with leach residue	2000 and 2001 programs	
Upgradability of rougher jig concentrate	With J1800 concentrate using J200 jig	With J200 concentrate using Mozley Laboratory Separator

yields a narrowly sized concentrate, ideal as spiral feed. Finally, the actual feed of the J1800 was not available in large quantities, and much of the piloting was completed on material being processed at Granny Smith at the time.

To mitigate the risk associated with the project, test work was extensive. Table 5 summarizes this work, paralleling the large-scale studies with the existing leach residues and smaller-scale studies with actual Wallaby residues.

CONCLUSIONS

The current generation of high-speed centrifugal concentrators offer new opportunities for fine heavy mineral recovery at a larger scale. Initial operating results from the Kelsey J1800 centrifugal jigs at the Granny Smith mine indicate that target heavy mineral and gold recoveries can readily be achieved. Optimisation studies will be conducted with the aim of increasing unit throughput while maintaining selectivity, measures which will increase the scope for application of these machines within the gold mining industry.

ACKNOWLEDGMENTS

We would like to thank the management of the Granny Smith Mine for the opportunity to present this paper and to acknowledge their support of and commitment to this new generation of gravity concentration technology.

REFERENCES

Beniuk, V.G., C.A. Vadeikis and J.N. Enraght-Moony (1994), "Centrifugal Jigging of Gravity Concentrates and Tailings at Renison Limited," in *Minerals Engineering,* Vol. 7(5/6). 577–589.

Laplante, A.R., (1984), "Sampling and Mass Balancing of Gold Ores," in *1st Intern. Symposium on Precious Metals Recovery,* Reno, Paper V, 15 p.

Martins, V., R. Dunne and G. Delahey, (1993), "New Celebration Tailings Treatment Plant—18 Months Later," in *XVIII International Mineral Processing Congress,* Sydney, 1215–1222.

Wyslouzil, H. (1990), "Evaluation of the Kelsey Centrifugal Jig at Rio Kemptville Tin," in *22nd Annual Meeting of Canadian Mineral Processor,* Ottawa, 461–472.

Applications of the HydroFloat Air-assisted Gravity Separator

Michael J. Mankosa[*], Jaisen N. Kohmuench[*], Graeme Strathdee[†], and Gerald H. Luttrell[‡]

Froth flotation is widely considered the most cost effective and versatile process for the separation of fine particles. Both column and conventional flotation have been successfully applied for the recovery of particles ranging from 30 mesh to sub-micron sizes. Unfortunately, the commercial application of flotation to recover coarser particles is not practical due to limits associated with bubble-particle buoyancy and detachment. These limits have been overcome through the development of an innovative air-assisted gravity separator known as the HydroFloat. The HydroFloat combines the versatility and selectivity of a flotation process with the low cost and high capacity of a gravity concentration process. This paper presents a review of the theoretical basis for this novel technology as well as data from phosphate, potash, feldspar, and other industrial mineral applications.

INTRODUCTION

Hindered-bed separators are commonly used in the minerals industry for particle classification. These units can also be employed for mineral concentration provided that the particle size range and density differences are within acceptable limits. However, these separators often suffer from the misplacement of low-density coarse particles to the high-density underflow. This shortcoming is due to the accumulation of coarse low-density particles that gather at the top of the teeter bed. These particles are too light to penetrate the teeter bed, but are too heavy to be carried by the rising water into the overflow launder. These particles are eventually forced by mass action downward to

---

[*] Eriez, Erie, Pa.

[†] PCS Potash, Saskatoon, Saskatchewan, Canada.

[‡] Dept. of Mining and Minerals Engineering, Virginia Polytechnic Institute and State University, Blacksburg, Va.

166 | NON-COAL GRAVITY SEPARATIONS

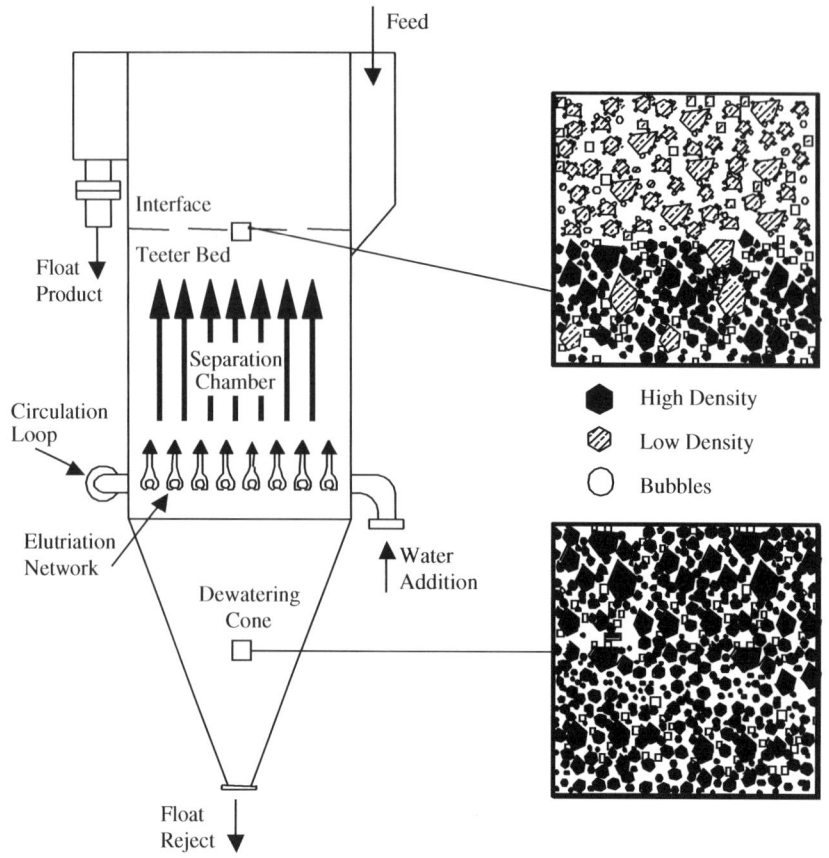

FIGURE 1 Schematic of the HydroFloat separator

the discharge as more particles accumulate at the top of the teeter bed. This inherent inefficiency can be partially corrected by increasing the teeter-water velocity to convey the coarse, low-density solids to the overflow. Unfortunately, the higher water rates will cause fine, high-density solids to be misplaced to the overflow launder, thereby reducing the separation efficiency.

To overcome the shortcomings of traditional hindered-bed separators, a novel device known as the HydroFloat separator was developed. As shown in Figure 1, the HydroFloat unit consists of a rectangular tank subdivided into an upper separation chamber and a lower dewatering cone. The device operates much like a traditional hindered-bed separator with the feed settling against an upward current of fluidization water. The fluidization (teeter) water is supplied through a network of pipes that extend across the bottom of the entire cross-sectional area of the separation chamber. However, in the case of the HydroFloat separator, the teeter bed is continuously aerated by injecting compressed air and a small amount of frothing agent into the fluidization water. The gas is dispersed into small air bubbles by circulating the water through a high-shear mixer in a closed-loop configuration with a centrifugal pump. The air bubbles become attached to the hydrophobic particles within the teeter bed, thereby reducing their effective density. The particles may be naturally hydrophobic or made hydrophobic through the addition of flotation collectors. The lighter bubble-particle aggregates rise to the top of the denser teeter bed and overflow the top of the separation chamber. Unlike flotation, the bubble-particle agglomerates do not need to have sufficient buoyancy to rise to the top of the cell. Instead, the teetering effect of the hindered bed forces the low-density agglomerates to overflow into the product launder. Hydrophilic particles that do not attach to the air bubbles continue to move down through the teeter bed and eventually settle into the dewatering cone. These particles are discharged as a high solids stream (e.g., 75% solids) through a control valve at the bottom of the separator. The valve is actuated in response to a control signal provided by a pressure transducer mounted to the side of the separation chamber. This configuration allows a constant effective density to be maintained within the teeter bed.

The HydroFloat separator can be theoretically applied to any system where differences in apparent density can be created by the selective attachment of air bubbles. Although not a requirement, the preferred mode of operation would be to make the low-density component hydrophobic so that the greatest difference in specific gravity would be achieved. Compared to traditional froth flotation processes, the HydroFloat separator offers several important advantages for treating coarse material. These include enhanced bubble-particle contacting, better control of particle residence time, lower axial mixing/cell turbulence, and reduced air consumption.

PROCESS THEORY

The reaction, or flotation, rate for a process is indicative of the speed at which the separation will proceed. In mineral flotation the reaction rate is controlled by several probabilities, e.g., collision, adhesion and detachment. The attachment of particles to air bubbles is the underlying principle upon which all flotation processes are based. This phenomenon takes place via bubble-particle collision followed by the selective attachment of hydrophobic particles to the bubble surface. Particles may detach if the resultant bubble-particle aggregate is thermodynamically unstable. According to Sutherland (1948), the attachment process may be described by a series of mathematical probabilities given by:

$$P = P_c P_a (1 - P_d) \quad \text{(EQ 1)}$$

in which P_c is the probability of collision, P_a the probability of adhesion, and P_d the probability of detachment. The attachment and detachment probabilities are controlled by the process surface chemistry and cell hydrodynamics, respectively. In an open (free settling) system, the collision probability is quite low due to the low particle concentration. However, at higher concentrations, the crowding effect within the hindered bed increases the probability of collision. This phenomenon is due to the compression of the fluid streamlines around the bubbles as they rise through the teeter bed. The increased probability of collision can result in reaction rates that are several orders of magnitude higher than found in conventional flotation.

Hindered-bed separators also operate as low-turbulence devices. As a result, particle detachment is minimized due to a reduction in localized turbulence. Studies conducted by Woodburn et al. (1971) suggest:

$$P_d = \left(\frac{D_p}{D_p^*}\right)^x \quad \text{(EQ 2)}$$

in which D_p is the particle diameter to be floated, D_p^* is the maximum floatable particle diameter, and x is an experimental constant (typically 3/2). Factors that influence the magnitude of D_p^* include pulp chemistry (surface tension and contact angle), physical particle properties (size, density, composition, and shape), and cell agitation intensity. Theoretical D_p^* values have been calculated by Schulze (1984) from the tensile and shear stresses acting on bubble-particle aggregates under homogenous turbulence. The degree of turbulence was quantified in terms of the induced root mean square velocity (*RMSV*). According to this theoretical study, the maximum size of particles that may be recovered by flotation increases by more than an order of magnitude when changing from high to low turbulence. According to Barbery (1984), the optimum conditions for coarse particle flotation occur when cell agitation intensity is reduced to a point just sufficient to maintain the particles in suspension. Thus, a teeter bed is an ideal environment for minimizing particle detachment.

The mixers-in-series model provides a convenient framework for analyzing this phenomenon (Arbiter and Harris, 1962; Bull, 1966). According to this model, the cumulative fractional recovery (*R*) of a given particle species can be determined using the expression:

$$R = 1 - (1 + k\tau_p)^{-n} \tag{EQ 3}$$

in which k is the flotation rate constant, τ_p is the particle residence time, and n is the number of equivalent mixers. Figure 2 shows recovery determined from the above relationship for different values of n as a function of the dimensionless product $k\tau_p$. In most cases, n is assumed to be equal to the number of cells in the flotation bank. This assumption is generally valid for a cell-to-cell flotation bank. However, the magnitude of n is typically smaller for flow-through flotation banks that have a significant amount of intermixing. The appropriate value of n can be readily estimated for any cell configuration using residence time distribution (RTD) data that have been collected using solid or liquid tracers. Details related to this procedure have been described elsewhere (Mankosa et al., 1992).

The HydroFloat cell operates under nearly plug-flow conditions due to the low degree of axial mixing afforded by the uniform distribution of particles across the teeter bed. As a result, the cell operates as if it were comprised of a large number of cells in series (i.e., high value of n). As shown in Figure 2, this characteristic allows a single unit to achieve the same recovery as a multi-cell bank of conventional cells (all other conditions equal). In other words, the HydroFloat cell makes more effective use of the available cell volume than well-mixed conventional cells or open columns.

The hindered-bed environment also influences particle retention time, and hence, particle recovery. In column flotation, particles settle vertically through the cell either with the fluid flow (co-current) or opposite to it (counter-current). A counter-current arrangement has obvious advantages since the settling velocity is reduced by the upward flow of liquid resulting in a higher retention time. Hindered settling, as previously explained, provides an environment in which the particles never achieve their terminal free-fall velocity. As a result, the effective particle velocity through the cell is greatly reduced, providing a significant increase in retention time as compared to a free-settling system. The longer retention time also allows good recoveries to be maintained without increasing cell volume.

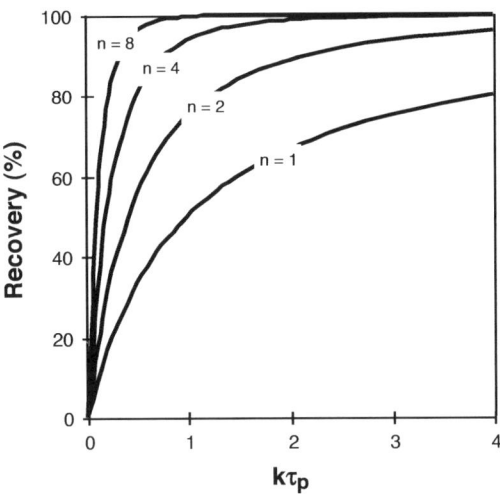

FIGURE 2 Theoretical recovery as a function of flotation rate and retention time

APPLICATION TESTING

Phosphate

The United States is the world's largest producer of phosphate rock. In 1999, this industry accounted for approximately 45 million tons of marketable product valued at more than $1.1 billion annually (United States Geological Survey, Mineral Commodity Summaries, January 1999). Approximately 83% of this production can be attributed to mines located in Florida and North Carolina.

Prior to marketing, the run-of-mine phosphate matrix must be upgraded to separate the valuable phosphate grains from other impurities. The first stage of processing involves screening to recover a coarse (plus 14 mesh) high-grade pebble product. The screen underflow is subsequently deslimed at 150 mesh to remove fine clays. Although 20–30% of the phosphate contained in the matrix is present in the fine fraction, technologies currently do not exist that permit this material to be recovered in a cost-effective manner. The remaining 14 × 150 mesh fraction is classified into coarse (e.g., 14 × 35 mesh) and fine (e.g., 35 × 150 mesh) fractions that are upgraded using conventional flotation machines, column flotation cells, or other novel techniques such as belt flotation (Moudgil and Gupta, 1989). The fine fraction (35 × 150 mesh) generally responds well to froth flotation. In most cases, conventional (mechanical) flotation cells can be used to produce acceptable concentrate grades with recoveries in excess of 90%. On the other hand, high recoveries are often difficult to maintain for the coarser (14 × 35 mesh) fraction.

Prior work has shown that the recovery of coarse particles (e.g., >30 mesh) can be less than 50% in many industrial operations (Davis and Hood, 1992). For example, Figure 3 illustrates the sharp reduction in recovery as particle size increases from 0.1 mm (150 mesh) to 1 mm (16 mesh) for a Florida phosphate operation. In many cases, attempts by plant operators to improve coarse particle recovery often produce an undesirable side effect of diminishing flotation selectivity.

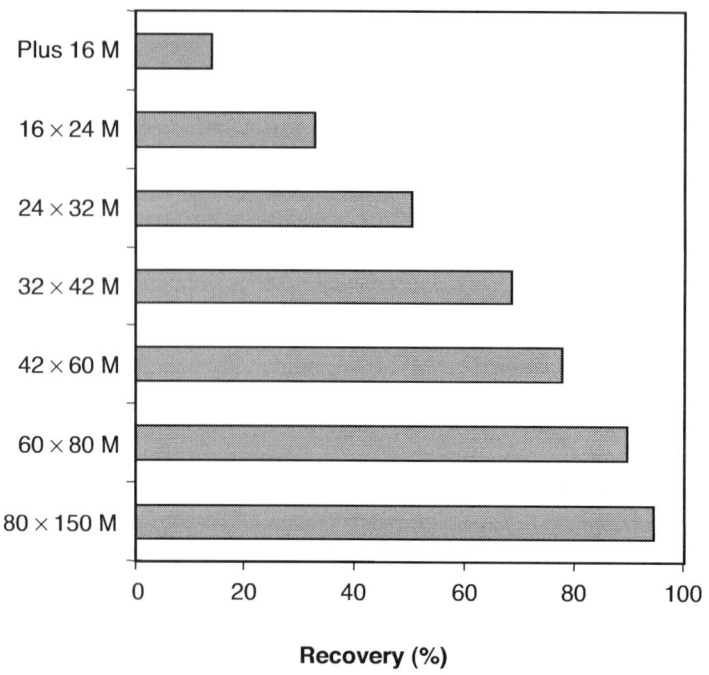

FIGURE 3 Size-by-size recovery for a typical North American phosphate producer

The United States Bureau of Mines (USBM) conducted one of the most comprehensive studies of the coarse particle recovery problem in the phosphate industry (Davis and Hood, 1993). This investigation involved the sampling of seven Florida phosphate operations to identify sources of phosphate losses that occur during beneficiation. According to this field survey, approximately 50 million tons of flotation tailings are discarded each year in the phosphate industry. Although the tailings contain only 4% of the matrix phosphate, more than half of the potentially recoverable phosphate in the tailings is concentrated in the plus 30 mesh fraction. In all seven plants, the coarse fraction was higher in grade than overall feed to the flotation circuits. In some cases, the grade of the plus 30 mesh fraction in the tailings approached 20% P_2O_5. The USBM study indicated that the flotation recovery of the plus 35 mesh fraction averaged only 60% for the seven sites included in the survey. Furthermore, the study concluded that of the seven phosphate operations, none have been successful in efficiently recovering the coarse phosphate particles.

Based on the established needs of the phosphate industry for a coarse particle recovery system, a pilot-scale test program was undertaken to evaluate the HydroFloat separator at a major phosphate plant in Florida. A schematic of the pilot-scale test circuit used for this study is shown in Figure 4. The test circuit consisted of three primary unit operations: pilot-scale classifier, slurry conditioner, and HydroFloat separator. In this circuit, the coarse underflow from an existing bank of classifying cyclones was fed to a 1.5 × 1.5 meter (5 × 5 ft) teeter-bed classifier (Eriez CrossFlow Separator). The underflow from the classifier was passed to the conditioning unit where appropriate reagents were added to control pH (ammonia) and particle hydrophobicity (fatty acid/fuel oil blend).

The test circuit was configured so that feed conditioning could be performed using either a stirred-tank (four stage) or a single-stage rotary drum (76-cm/30-inch diameter)

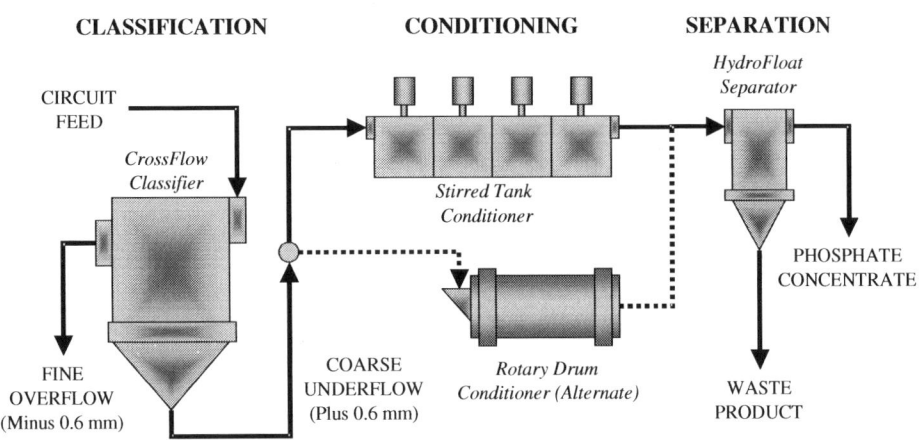

FIGURE 4 In-plant pilot-scale test circuit

conditioner. The conditioner circuit was able to operate reliably at approximately 40–75% solids at a maximum mass flow rate of 4–6 tonne/hr (dry solids). The conditioned slurry flowed by gravity to the feed inlet for either the HydroFloat separator or a 50-cm (20-inch) diameter flotation column. This arrangement made it possible to directly compare the effectiveness of the HydroFloat separator with existing column technology. The test circuit was installed with all necessary components (i.e., reagent pumps, etc.) required to operate the separator in continuous mode at the desired capacity.

To compare the HydroFloat with current state-of-the-art column technology, comparison tests were conducted with an open-column flotation cell. The column utilized state-of-the-art sparger technology and was supplied with instrumentation to maintain level and monitor air and water flow rates. Comparison tests were conducted on each cell as a function of various operating conditions. The objective of the test program was to collect sufficient data using each separator and generate comparable product grade versus recovery curves.

The results from the column comparison tests are presented in Figures 5 and 6. The data shown in Figure 5 indicate that both the HydroFloat and open column operated on the same product grade versus recovery curve. The BPL recovery, however, was substantially higher for the HydroFloat system. The result is particularly impressive considering that the open column was operated at a substantially lower feed rate than the HydroFloat. As shown, the open column was able to achieve BPL recoveries exceeding 90% at a feed rate of 6.5 tph/m^2 (0.66 tph/ft^2). However, as the feed rate increased to a higher value of 9.8 tph/m^2 (1.0 tph/ft^2), the BPL recovery dropped significantly. The HydroFloat, on the other hand (Figure 6), was able to maintain a BPL recovery averaging 98% at a feed rate exceeding 19.6 tph/m^2 (2.0 tph/ft^2). It should be noted that at a feed rate of 24.5 tph/m^2 (2.5 tph/ft^2), the capacity of the conditioner (not the HydroFloat) was exceeded. At this capacity, poor conditioning caused a decrease in the downstream performance of the HydroFloat separator. These results clearly demonstrate that the HydroFloat capacity exceeds that of flotation column cells currently used by the phosphate industry.

A summary of all of the test data obtained during this evaluation is provided in Figure 7. The improved flotation response for the drum conditioner, which was demonstrated in an earlier laboratory evaluation, was also verified through the pilot-scale testing. As shown, the rotary conditioner improved BPL recovery by approximately 20 percentage

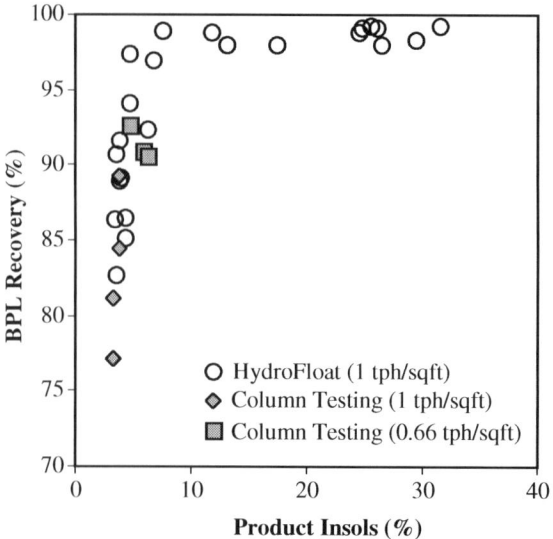

FIGURE 5 BPL recovery comparison for the column and HydroFloat separators

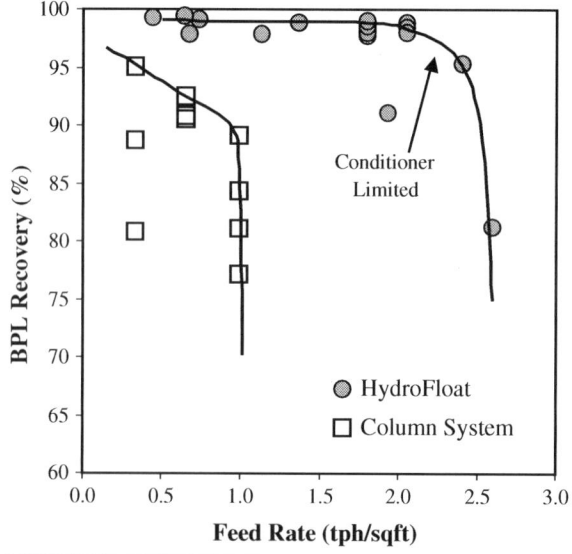

FIGURE 6 Feed rate comparison for the column and HydroFloat systems

points. In fact, the BPL recovery approached 98% at a product insols grade between 5% and 12%. The high concentrate grade is due to the improved recovery of the coarse, high-grade particles normally lost in traditional mechanical flotation. When using the Hydro-Float system, over 80% of the coarsest phosphate particles (+10 mesh) were recovered. Figure 8 shows the typical size-by-size BPL recoveries and insols rejections obtained using the HydroFloat separator. As shown, the HydroFloat was able to maintain a high BPL recovery and insols rejection for all size classes.

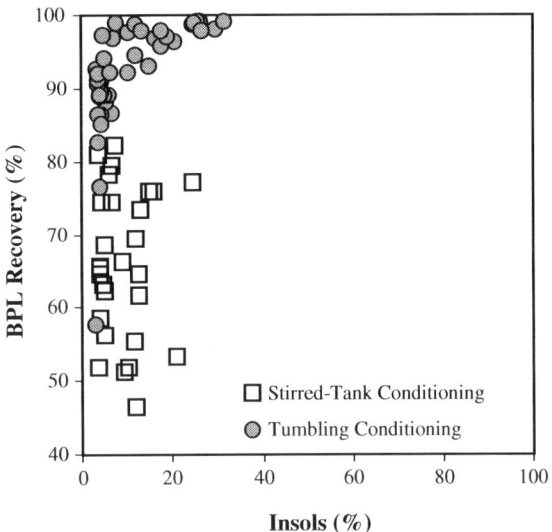

FIGURE 7 Overall BPL recovery versus product insols curve

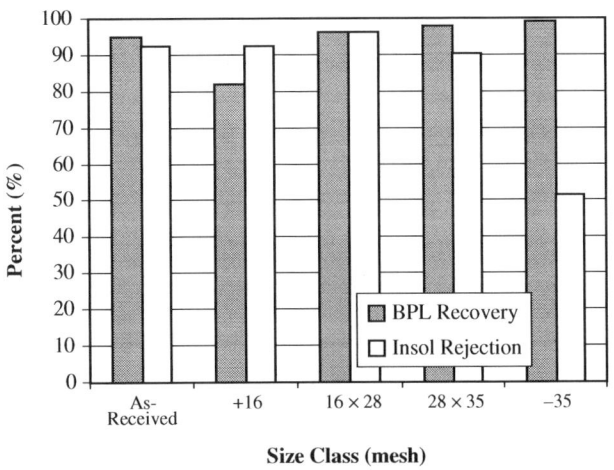

FIGURE 8 Typical size-by-size recovery for in-plant HydroFloat evaluation

Potash

An important commodity for the manufacture of fertilizer, potash occurs as sylvinite ores that contain sylvite (KCl), halite (NaCl), and insolubles. Flotation of the sylvite is achieved using a cationic collector. Flotation is carried out in saturated brine using standard conventional flotation machines. Much like phosphate, potash is also recovered at a coarse particle size. The coarse concentrate commands a premium price and eliminates the need for granulation and compaction (Soto, 1993). With a top size of approximately 5 mesh, however, recovery of the coarse material is often low in comparison to the fine fraction (i.e., <1 mm). In an attempt to improve coarse particle

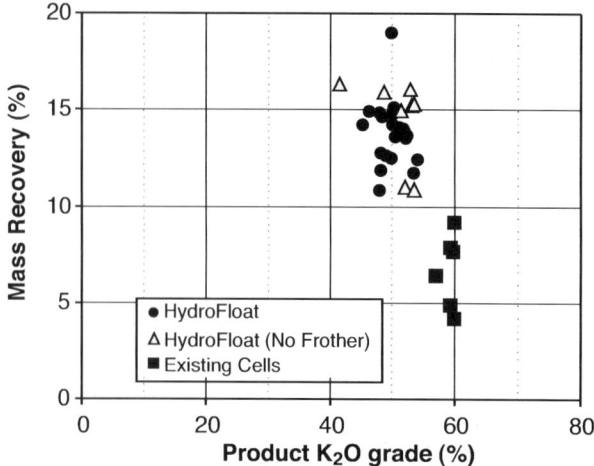

FIGURE 9 Mass recovery versus product grade for HydroFloat in-plant testing of potash

recovery, an in-plant evaluation of the HydroFloat was conducted at PCS Potash in Saskatchewan, Canada.

Tests were conducted on a tailings stream from existing rougher flotation cells. This coarse reject fraction (mean size = 2.2 mm) contains approximately 8% K_2O and is currently being upgraded in a bank of conventional scavenger cells. Conditioning of the ore is completed at high percent solids (+75%) using an amine collector. A sample of this material was fed continuously to a 30-cm (12-inch) diameter, pilot-scale HydroFloat separator. Process brine was used as the fluidizing medium.

Tests were conducted as a function of air and elutriation rate in order to produce a grade versus recovery curve for the potash ore. For each test, samples of the product, reject, and feed streams were collected for assay. The solids feed rate to the HydroFloat was maintained at 15.7 tph/m^2 (1.6 tph/ft^2) and was delivered at 70% solids. For comparison, samples were also collected from the existing conventional cells that were operating in parallel to the HydroFloat.

The results from the in-plant tests are shown in Figures 9 and 10 as K_2O and mass recovery versus product grade. As shown, the average recovery of K_2O using the HydroFloat system is roughly double that which can be achieved by conventional flotation cells. The average K_2O recovery for the HydroFloat was 90.4% while the existing cells averaged 51.9% recovery. Unlike the HydroFloat, mechanical cells have a relatively short retention time, a high degree of turbulence (mixing), and a low probability of bubble/particle contacting due to the low percent solids content of the pulp. The obvious advantage of the high recovery offered by the HydroFloat system is the overall increase in mass yield or product tons. In fact, the average mass yield for the existing cells was 6.7%, while the HydroFloat achieved an average 13.9% weight recovery.

Feldspar

Feldspar is mined in several locations throughout the world and is used predominantly in the glass and ceramics industries. Difficulties are often observed with regard to coarse particle recovery via flotation. Consequently, feldspar ore is typically ground much finer than the liberation size in order to optimize plant flotation circuits. Mica

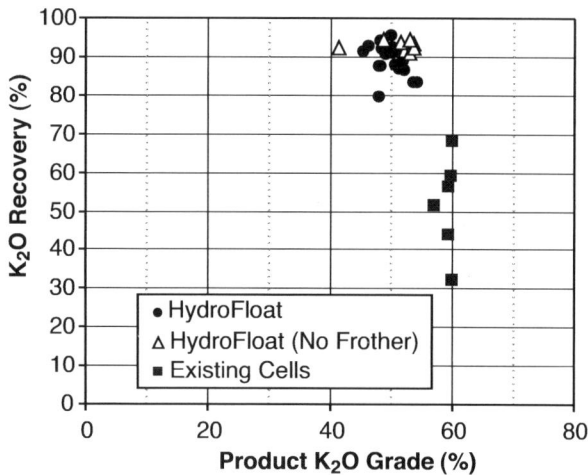

FIGURE 10 K_2O recovery versus product grade for HydroFloat in-plant testing of potash

and iron-bearing minerals are also found in the raw ore and are initially removed by flotation prior to the final separation of the high-grade feldspar particles from the silica gangue. Regardless, the successful recovery of coarse feldspar yields many advantages including the reduction of grinding costs and the subsequent decrease in fines production (i.e., over-grinding).

To evaluate the effectiveness of the HydroFloat for coarse feldspar recovery, a series of laboratory-scale tests were conducted on a bulk sample obtained from an existing producer. The starting sample was nominally 20 × 100 mesh and had a feldspar content of 25%. The predominant gangue mineral was quartz. The separation objective was to achieve a product containing 65% feldspar at a recovery of at least 85%. Tests were conducted as a function of feed rate (9.8–14.6 tph/m² or 1.0–1.5 tph/sqft) and collector addition. The process water was adjusted to a pH of 2.5 using sulfuric acid and conditioning was accomplished using a tumbling-type conditioner. Aeration rate was maintained at 0.3 m³/min/m² (1 scfm/sqft). Samples of the feed, product, and reject streams were collected from each test for analysis. The results from this series of tests are shown in Figure 11 as feldspar grade versus recovery. These findings indicate that the Hydro-Float can recover 90% of the feldspar mineral at a concentrate grade greater than 65%.

Coal

Gravity concentration is used almost exclusively in the coal industry for upgrading coal to achieve the desired market product specifications. In most applications, the run-of-mine coal is classified into three or more size fractions prior to separation. The smallest size fractions are typically 28 × 100 and –100 mesh. The –100 mesh fraction is treated using froth flotation while the 28 × 100 mesh material is usually concentrated with spirals. Spirals are flowing-film, gravity concentrators that effectively separate "sand size" material according to specific gravity.

More recently, teeter-bed separators are also being implemented for concentration of the 28 × 100 mesh fraction. Teeter-bed separators offer many advantages including less floor space, less head room, and automatic control. Additionally, teeter bed separators have a unit capacity up to 24.5 tph/m² (2.5 tph/sqft). This translates to 230 tph for

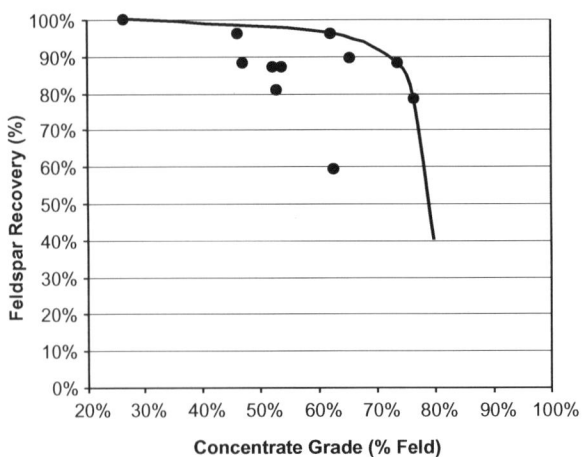

FIGURE 11 Feldspar product grade versus recovery

a single 3 × 3 meter (10 × 10 ft) separator. By comparison, spirals have a unit capacity of 3–3.5 tph. Therefore, multiple spirals must be used to process typical plant tonnages. As a result, the performance of a spiral circuit must be quantified by averaging the efficiency of all the individual separators. In the previous example, this implies that 72 spirals must operate at the same process efficiency.

From a process efficiency point-of-view, the advantages of a single-unit operation are obvious. Unfortunately, teeter-bed separators can also suffer from process inefficiencies if oversize material is present. As described elsewhere (Reed et al., 1995; Honaker, 1996), a teeter-bed separator classifies material based on the terminal hindered settling velocity. Fine and low-density particles report to the overflow while coarse and high-density material discharge through the underflow. As a result, misplaced coarse, low-density material will be lost with the tailings, resulting in a lower process efficiency.

The HydroFloat separator overcomes this problem through the addition of a small amount of air with the teeter water. In coal applications, the naturally hydrophobic particles adhere to the air bubbles reducing the effective density of the coarse particles. As a result, coarse coal particles that were previously lost to tailings are forced to the overflow with the clean coal concentrate. It should be noted that this is not a flotation process. Without the upward fluidization water, the bubble-particle aggregate will not have sufficient buoyancy to rise to the top of the separator. The bubble-particle aggregate behaves as a single entity with a lower specific gravity.

To illustrate the advantage of the HydroFloat in fine coal recovery, a series of in-plant tests were conducted on a central Appalachian coal. Tests were conducted using a 30-cm (12-inch) diameter pilot-scale separator at feed rates ranging from 15–25 tph/m^2 (1.5–2.5 tph/sqft). In this evaluation, the HydroFloat was operated with and without air injection. In the absence of air, the HydroFloat operates as a teeter-bed separator. The results from these tests are shown in Figure 12. It can be seen that the HydroFloat is able to achieve a higher product mass yield at the same product quality (ash content). The improvement is attributed to increased recovery of the

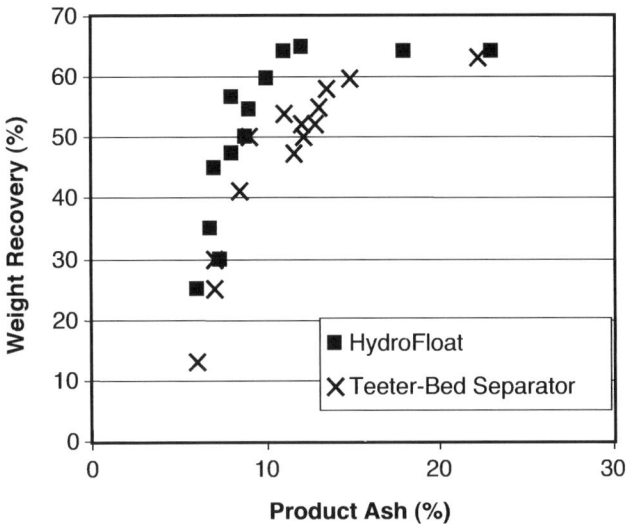

FIGURE 12 Clean coal mass recovery versus product ash content for teeter bed and HydroFloat separators

coarse, low-density coal particles. In fact, washability analysis of the separated products indicated that the HydroFloat recovered approximately 30 percent more of the +1.6 s.g. particles in the +35 mesh size fraction.

CONCLUSIONS

A new separator, known as the HydroFloat, has been developed to overcome some of the shortcomings associated with traditional separation processes for recovering coarse particles. The novel characteristic of this separator is the formation of a hindered "teeter" bed of fluidized solids into which small air bubbles are introduced. The bubbles attach to hydrophobic particles and create light bubble-particle aggregates that can be separated from hydrophilic particles based on the principle of differential density. Benefits of this new separator include enhanced bubble-particle contacting, better control of particle residence time, lower axial mixing/cell turbulence, and reduced air consumption.

Laboratory tests were conducted to evaluate the potential of this new technology for upgrading mineral samples from various sources (e.g., phosphate matrix, potash, feldspar, and coal). The test data indicate that the HydroFloat cell is capable of increasing coarse particle recoveries by 20% over conventional separation approaches. Furthermore, the concentrate grades were also improved in some cases due to a reduction in coarse particle misplacement.

Based on the success of the laboratory test work, pilot studies were undertaken for three specific applications: phosphate, potash, and coal. In each case, tests were conducted using larger scale separators capable of processing up to 6 tph. The results from each application indicate that the HydroFloat can achieve a substantially higher recovery at the same product grade as compared to current plant performance. A summary of the findings from these studies is presented in Table 1.

TABLE 1 Summary of results from HydroFloat plant studies

Application	Feed Rate (tph)		Recovery (%)		Grade (%)	
	Current	HydroFloat	Existing	HydroFloat	Existing	HydroFloat
Phosphate	10 tph/m^2	25 tph/m^2	91	98	4–6[1]	4–6[1]
Potash	7 tph/m^2	16 tph/m^2	50	90	60[2]	57[2]
Coal	3 tph/start	20 tph/m^2	50	60	8–10[3]	8–10[3]

(1) Percent insols; (2) % K$_2$O; (3) % ash.

REFERENCES

Arbiter, N. and Harris, C.C., 1962. "Flotation Kinetics," *Froth Flotation–50th Anniversary Volume,* (D.W. Fuerstenau, Ed.), Chpt. 8, American Institute of Mining, Metallurgy and Petroleum Engineers, Inc., New York, NY, pp. 215–246.

Barbery, G., 1984. "Engineering Aspects of Flotation in the Minerals Industry: Flotation Machines, Circuits and Their Simulation," *The Scientific Basis of Flotation,* (K. J. Ives, Ed.), NATO Advanced Institute Services, Series E: Applied Sciences, No. 25, Martinus Nijhoff Publishers, Boston, MA, pp. 289–348.

Bull, W.R., 1966. "The Rates of Flotation of Mineral Particles in Sulphide Ores," *Proceedings,* Australian Institute of Mining and Metallurgy, No. 220, pp. 69–78.

Davis, B.E. and Hood, G.D., 1992. "Conditioning Parameter Effects on the Recovery of Coarse Phosphate," *Proceedings,* Regional AIME/AIChE/AIPG/FIPR Phosphate Conference, Lakeland, FL, Sept. 24–25.

Davis, B.E. and Hood, G.D., 1993. "Improved Recovery of Coarse Florida Phosphate," *Mining Engineering,* Vol. 45, No. 6, pp. 596–599.

Honaker, R.Q., 1996. "Hindered Bed Classifiers for Fine Coal Cleaning," *Proceedings,* 13th International Coal Preparation Conference, Lexington, KY, pp. 59–70.

Mankosa, M.J., Luttrell, G.H., Adel, G.T., and Yoon, R.-H., 1992. "A Study of Axial Mixing in Column Flotation," *International Journal of Mineral Processing,* Vol. 35, pp. 51–64.

Moudgil, B.M. and Gupta, D., 1989. "Flotation of Coarse Phosphate Particles," Advances in Coal and Mineral Processing Using Flotation, *Proceedings,* Engineering Foundation Conference, Dec. 3–8, pp. 164–168.

Reed, S., Roger, R., Honaker, R.Q., and Mankosa, M.J., 1995. "In-Plant Testing of the Floatex Density Separator for Fine Coal Cleaning," *Proceedings,* 12th International Coal Preparation Conference, Lexington, KY, pp. 163–174.

Soto, H., 1993. "Pilot Testing of Column for Coarse Particles Flotation," SME Annual Meeting, Reno, NV, *Preprint,* Number 93–141.

Schulze, H.J., 1984. "Physico-Chemical Elementary Processes in Flotation," *Developments in Mineral Processing,* Vol. 4, Chpt. 5, Elsevier, NY, pp. 238–253.

Sutherland, K.L., 1948. "Kinetics of the Flotation Process," *Journal of Physical Chemistry,* Vol. 52, p. 394.

United States Geological Survey, Mineral Commodity Summary, January 1999.

Woodburn, E.T., King, R.P., and Colborn, R.P., 1971. "The Effect of Particle Size Distribution on the Performance of a Phosphate Flotation Process," *Metallurgical Transactions,* Vol. 2, pp. 354–362.

Advances in the Application of Spiral Concentrators for Production of Glass Sand

Steve Hearn[*] and Jim Sadowski[*]

Around the world, spiral concentrators have been successfully applied to glass sand production. Spiral application is both cost effective and environmentally friendly when compared to other techniques for iron bearing and refractory heavy mineral rejection such as flotation and magnetic separation. The coupling of spiral concentrators with hydraulic density separators for damp sand production often results in a process that meets both particle size and mineral/chemistry specifications. Interestingly, glass sand size specifications usually correlate directly to the optimum response particle size for spiral concentrator operation. This paper presents flowsheet alternatives and resulting process performance for glass sand operations representative of commercial operations around the world. The paper also suggests the use of spiral concentrators for rejection of aluminum silicates and mica from quartz sand destined for glass making markets.

INTRODUCTION

The Role of Spiral Concentrators in Glass Sand Production

Glass sand production is dependent on the mineral occurrence characteristics in the deposit, and most importantly, market requirements. Spiral concentrators have long played an important role in the production of saleable glass sand at sites around the world. The traditional purpose of the spiral units is removal of liberated heavy iron-bearing minerals from the sand. Of course, those iron-bearing minerals that are fully liberated will easily be rejected in the spiral, whereas silica sand grains with minor inclusions of the contaminant iron mineral will not reject. Liberation to a degree is imperative for successful application of the spiral technology. With proper feed preparation, i.e., density separator sizing, improved performance capabilities have been realized.

---

[*] Outokumpu Technology, Inc., Jacksonville, Fla.

The unwanted minerals in glass sand for the most part are iron-bearing minerals. These minerals have significantly higher specific gravities than quartz. Contaminant minerals such as magnetite and ilmenite, for example, have specific gravities of 4.0 or higher compared to quartz at 2.65. Typically, spirals can separate minerals with a specific gravity differential greater than 0.5 units with high efficiency, which makes this separation relatively easy.

In addition to the iron-bearing minerals, aluminum-bearing minerals, such as refractory aluminum silicates and mica, are also likely candidates for rejection in a spiral. The separation of these minerals from quartz is more difficult and requires a slightly different approach than removal of the iron minerals. Testwork has shown promise for removal of mica in a wash water assisted spiral.

Spiral concentrators offer a relatively simple unit operation that translates to low capital and operating cost. This, coupled with reagent free processing, provides the necessary low cost and environmentally desirable process. The actual spiral plant flowsheet most suitable for a particular application will depend on the feed characteristics, especially particle size distribution and mineralogical characteristics of the resource. However, certain generalizations apply. For instance, in the production of a marketable quality sand is most always the primary driver, with weight recovery of secondary importance. Therefore, two-stage separation is often advisable to ensure final sand product quality.

Spiral Concentrator Design and Operating Basics

Spiral manufacturers now offer a variety of models to the industry, each with specific helix trough profile designs, pitches, and other performance improvement nuances. These lightweight models are made of urethane-lined fiberglass and can be expected to last in excess of 10 years even under heavy service conditions. Compared to other sand beneficiation process such as flotation and magnetic separation, spirals present a relatively low capital cost, have no moving parts and consequently have low maintenance costs. Compared to dry magnetic separation, the spiral process is conducted on wet material and therefore the product does not require drying. In those parts of the world that can sell damp glass sand there is no need to expend the capital and energy cost to dry the sand. Even in places that require dry sand, the removal of contaminates prior to drying saves energy cost. If necessary, additional separation efficiency can be achieved by hydraulic classification of the feed and/or re-treatment of the first pass sand in a second pass through a spiral unit.

Spiral Feed Presentation. Spiral concentrators are flowing film separators that work in a similar principle to shaking tables. The design of the feed box is important to assure proper presentation of the feed slurry to the spiral trough resulting in desirable flow characteristics down the helix trough. With proper feed presentation, the separation process begins immediately at the top of the spiral helix. If the box design is problematic, i.e., presenting the spiral with an uneven or unbalanced feed, the pulp will have to stabilize in the trough before separation initiates and that can require up to one complete turn (revolution) of the helix, thus losing separation potential within the length of the spiral. In addition, if heavy minerals targeted for rejection via product cutters at the inner edge of the spiral trough, somehow through unnecessary turbulence, reach the high water (outer) portion of the trough, their ability to re-enter the flowing pulp and migrate to the center portion of the spiral is improbable and therefore, these particles will report to the glass sand product.

Heavy Mineral Entrapment. Another area of concern is entrapment. As the feed pulp flows down the spiral, heavy minerals can be trapped below the bed of sand in the middle portion of the spiral. Like a temperature inversion, these particles become trapped under the blanket of sand and are unlikely to migrate out and into the region of the heavy minerals at the inner area of the trough. To counteract this problem, spirals are often equipped with surface bumps or repulper designs that help free these trapped minerals and allow them to migrate and report to the proper location within the spiral.

Benefit of Centrifugal Force. As slurry flows down the spiral helix, there is a centrifugal force acting to push the lighter minerals up the trough profile to the outside region. The force is not sufficient to move the heavier minerals to this region therefore they slide down the profile to the inside region of the spiral.

Since this is a flowing film separator, the centrifugal forces are not equal along the depth of the slurry. At the very surface of the spiral, the centrifugal forces are very small. Therefore, the smaller particles have little forces acting on them. At higher depths of the slurry, the forces become greater and therefore, the coarser particles have higher forces acting on to help them report to the outer regions of the spiral.

For glass sands, this variation in forces between the finer and coarser particles generally is a benefit since the majority of the heavy minerals in the deposits tend to be finer than the glass sand. Therefore, the lower forces on the heavy mineral particle help to allow them their reporting to the center of the spiral for rejection. The higher forces on the coarser glass sand particles push them toward the outside and allow them to report to the glass sand product. In many cases with glass sand, the separation between the heavy iron bearing minerals and the sand portion is so extreme that the urethane surface of the spiral can actually be seen between the two particle streams flowing in the trough.

USING SPIRALS TO REMOVE IRON-BEARING HEAVY MINERALS

Throughout the world, there is considerable experience in the use of spirals to remove the iron bearing heavy minerals in glass sand deposits. There are currently more than 25 glass sand spiral installations around the world.

In these installations, spirals having either five or seven turn helixes are employed; the number of turns required depends on the amount of heavy iron-bearing minerals that need to be removed, i.e., with higher amounts of particles needing removal then the longer (7-turn) spirals are desirable. At times, there is also a benefit of passing rougher stage (1st pass) spiral sand product through a second stage of spirals (cleaner pass) to remove the remaining heavy minerals. For example, if the first stage is 70% effective then the second or cleaning stage will be nearly 70% effective on the remaining 70% heavy minerals. Thus, the two-stage spiral process can be said to be 91% effective in heavy mineral rejection.

The cleaner stage also fits well from a plant layout and operational standpoint. In the spiral process, the first stage removes a heavy mineral product at a high pulp density, from the inner area of the spiral trough helix. The lighter glass sand product along with most of the feed pulp water reports to the outer area of the trough helix where it exits the spiral. The pulp density of the first pass sand product is at a reasonable density for directly feeding the second stage. Therefore, the first and second stage spirals can be stacked one above the other with the product from the first stage reporting directly to the second stage. Using this practice eliminates the cost of pumping pulp from the first to the second stage. However, where height restrictions prevent stacking, traditional pumping between spiral stages can be employed.

Spiral separation typically can remove 60 to 80 percent of the heavy minerals in the feed stream. This is dependent on the amount and size distribution of the heavy minerals compared to the silica sand portion. Data from various operations are discussed below.

Example 1: Materiales del Istmo, Veracruz, Mexico

Successful implementation of a spiral processing circuit in place of flotation was realized in the mid nineties. A traditional flotation circuit operated on feed having a silica-to-feldspar ratio of 80:20. Their existing flotation plant consistently made a quality product. This required flotation reagents of up to 500 g/t (additional reagents were required for waste water treatment) and residual fatty acids and other flotation reagents in the plant effluent water produced difficult environmental problems for management.

Installation of a spiral based plant (Carpco® LC3000 spirals) after in-plant trials, resulted in an increase in production of more than 40% and an improved final product iron content (as Fe_2O_3) ranging from 0.027 to 0.038% compared to 0.045 to 0.070% from the flotation circuit. The elimination of flotation reagents improved wastewater treatment by eliminating the need for lime addition. Overall, operability of the plant improved with less power required and the simpler circuit required less operator attention. Significant cost savings were achieved.

Example 2: Silice del Istmo, Veracruz, Mexico

Another spiral plant producing a quality glass sand product in Mexico is the Silice del Istmo operation, also in Veracruz. In this operation the Carpco® LC3000 spirals are operated at nearly 1.8 t/ht/h per spiral start in a plant treating a total of 107 t/h of new feed. Interestingly, even at this relatively robust feed rate, the resulting quartz product presently contains 0.055% Fe_2O_3 produced from a feed of 0.10% Fe_2O_3. The nature of this circuitry results in a natural sand product size –30+140 mesh which is appropriate for the glass sand market.

Example 3: Meren, Jordan

Another silica sand plant that is slated for start-up in late 2003 utilizes a double stage LC3000 spiral circuit to improve the iron content. This particular deposit is much higher grade or lower iron than the previously sited examples. The plant has been designed to produce two grades of silica sand. The standard grade sand will contain less than 0.020% Fe_2O_3 and the premium grade sand will contain less than 0.012% Fe_2O_3 from a feed that contains from 0.030 to 0.017% Fe_2O_3. Both products will report to the spiral circuit with the premium sand also being dried and further processed with rare earth roll magnetic separators.

In the process, the feed reports to a Floatex® density separator prior to the spiral circuit. The main purpose of the Floatex® density separator is to remove the +0.6 µ (30 mesh) size fraction. However, the test showed an addition benefit of also removing a portion of the heavy minerals as shown in Table 1. The data is compared to sizing resulting from screen. Screening produces a product based on true size whereas the density separator also incorporates the specific gravity of the particles along with particle size.

Based on true sizing, there were no heavy minerals in the +600 µ fraction. However, with the density separator sizing, 21.7% of the heavy minerals reported to the coarse fraction. In the plant, the coarse fraction iron content in not important to the end-use. It

TABLE 1

Size (µ)	Density Separator % HM	HM % Dist		Screening % HM	HM % Dist
+600	0.07	21.7	+600	0	0
			−600+300	0.04	16
−600	0.11	72.9	−300	0.56	84

TABLE 2

Product	Wt%	% HM	HM % Dist
1st pass rejects	3.8	1.1	40.0
2nd pass rejects	4.5	0.91	30.0
Middling	8.0	0.06	5.0
Products	83.7	0.03	25.0

TABLE 3

Location	Feed % Fe_2O_3	Standard Product % Fe_2O_3	Premium Product % Fe_2O_3
Pit 1	0.024	0.015	0.010
Pit 2	0.019	0.012	0.010
Pit 3	0.028	0.020	0.012
Pit 4	0.030	0.019	0.010
Pit 5	0.017	0.015	0.009
Pit 6	0.024	0.013	0.012

should also be noted that as previously discussed, the spiral would have difficulty removing the coarse heavy minerals.

The −600 µ fraction from the above process then reported to the spiral circuit for additional heavy mineral rejection, as shown in Table 2.

The data shows that the spirals were very effective in removing the heavy minerals with 70% of the total heavy minerals reporting to 8.3 weight percent of the feed.

Glass sand however is not sold on the percentage of heavy minerals but rather the iron content of the final product. Table 3 shows the results of bulk tests conducted on samples from different areas of the proposed mine site.

The data show that the spiral circuit was able to meet the objectives of the customer with the average of the six bulk tests at 0.0156% Fe_2O_3 for the standard product.

OPPORTUNITY FOR SPIRAL REJECTION OF MICA

Mica separation is a difficult operation regardless of the process, although flotation and magnetic separation are somewhat effective. Complicating the use of flotation is that a different reagent scheme is often needed to remove mica compared to removal of the heavy iron-bearing minerals. Magnetic separation is effective but the feed rates are often very low since some mica is only weakly magnetic and is a relatively expensive operation, especially after drying.

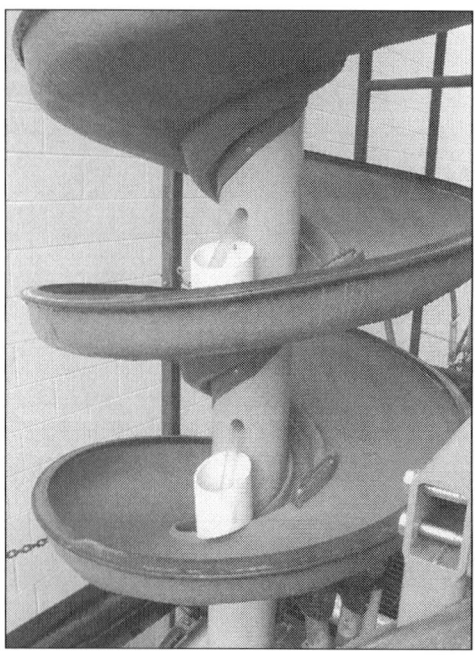

FIGURE 1 Carpco® CS2000W

Spirals have been used for primary mica recovery for many years, while at this time there are no commercial installations for its removal from quartz sand product. Outokumpu has recently introduced a new spiral that has shown good success in removing mica from glass sand. This spiral incorporates a wash water system in a shallow pitch spiral. The wash water aids in pushing the mica outward and into the high water region of the spiral. In conventional spiral operations, the mica becomes trapped at the base of the sand region and flows with the quartz sand to the spiral discharge. Although mica has a specific gravity similar to quartz, its shape factor can make its apparent specific gravity lower.

In this new spiral specifically designed for mica removal, the quartz sand material reports to the central part of the spiral (note that in conventional spiral application heavy minerals would report to this inner trough region). Trough splitters are used in the new spiral to direct the quartz sand product to the inner cutter at each of the spiral turns. Immediately after each splitter position, wash water is added to push (or flush) mica out of the sand fraction to the high water region of the trough for eventual rejection to tailing at the discharge of the spiral. Figure 1 illustrates the spiral separation mechanics.

Data have been developed through both laboratory testing and in-plant trials. Qualitatively, it has been observed that varying the amount of wash water can affect the amount of mica rejected, and similarly, the recovery of quartz. Table 4 shows a set of data from a pilot test of the new mica rejection spiral.

The data shows that in excess of 62% of the mica was rejected, while only 2–3% of the quartz was lost in the spiral process. The reject material contained 88.4% mica with a majority of the remaining material being fine quartz particle that would be ultimately rejected in other portions of the process. Therefore, in addition to the spiral being capable of removing the mica, the washing effect also removed some of the undesirable fine particles.

TABLE 4

Product	Wt%	% Mica	% Mica Dist
Feed	100	23.2	100.0
Product 1	51.4	11.5	24.7
Product 2	31.6	9.5	12.5
Product 1+2	83.0	10.7	37.5
Reject	17.0	88.4	62.8

In commercial operation, the quartz sand product is dried and then sent to dry magnetic separation where 20–25 wt% is rejected to the magnetic product which contains the undesirable mica. Assuming a drying rate of 50 t/h, there is a 10 t/h of rejected material that could be removed by the spiral prior to final product drying. A simple projection of annual savings in natural gas for drying this material equates to approximately $50,000 per annum. Economics may differ outside North America as sand is typically shipped damp elsewhere around the world.

SPIRAL REJECTION OF REFRACTORY HEAVY MINERALS

In addition to mica and mica-like aluminum bearing minerals, refractory aluminum silicate minerals also present a problem to glass manufacturers. Fortunately, there are large numbers of high quality glass sand resources around the world that are void of the refractory heavy minerals. However, in some locations such as India, the geology is such that it is a common occurrence to find high quality quartz sand contaminated with these refractory heavy minerals. Their presence in even small amounts such as a few grains per kg of sand can cause glass producers considerable problems since each grain likely results in a glass melt defect.

The refractory heavy minerals pose a unique problem for spiral separation since the specific gravity of these minerals are close to that of quartz than are the iron-based heavy minerals. In addition, the grains that cause the most problems to the glass producers are the coarser grains since these are the most difficult to melt. Refractory mineral grains that are less than 200 μ do not pose much of a problem; however, those that are greater than 400 μ are very problematic.

As previously discussed, the coarser particles tend to be more difficult to separate in a spiral due to the higher centrifugal forces imposed on these grains in a spiral. These forces tend to push the coarse grains away from the center portion of the spiral and into the region of the finer quartz. The solution to this problem is to present the spiral with a narrower size distribution compared to a typical glass sand size distribution. When the spiral feed is of a narrow size distribution there is a better chance of rejection of the nearer gravity refractory minerals. Note that all gravity concentrators tend to confuse particle size with specific gravity differential and when it is possible to classify the feed into narrow size ranges, the resulting gravity based separation is improved.

Test work and industrial installations have shown that the best method to produce the narrow size distribution is by use of a density separator, sometimes referred to as a hydraulic classifier. These units, such as the Floatex® density separator, are more effective in front of gravity separation systems than screens when applied to making size cuts in the 300–500 μ range. These same density separators provide an additional processing benefit because of the inherent preferential sizing based on mineral specific gravity. For instance, a 400 μ refractory heavy mineral grain will report to the separator underflow

TABLE 5

Size μ	Size Mesh	Sillmanite (3.2 sg)	Olivine (3.5 sg)	Ilmenite (4.7 sg)
850	20	560	470	290
600	30	395	330	200
425	40	280	235	145
300	50	195	165	100
212	70	140	115	75
150	100	100	885	50
100	150	65	55	35
75	200	50	41	25

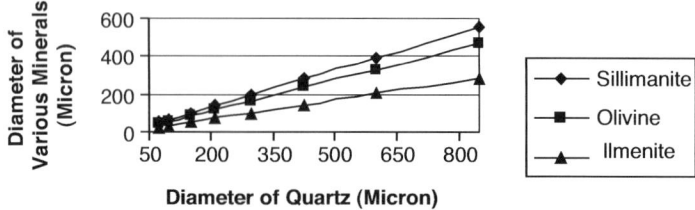

FIGURE 2 Size separation equivalent for various minerals versus quartz

(coarse product) whereas a 600 μ quartz grain will report to the separator overflow (fine product). The differential in sizing of various minerals compared to quartz is shown in the Table 5. The same data is then presented in graph form in Figure 2.

In a properly operated density separator, making a 600 μ size separation based on quartz, the equivalent size separation for sillmanite would be 395 μ, olivine 330 μ and ilmenite 200 μ. That is, quartz particles coarser the 600 μ would report to the underflow and finer than 600 μ to the overflow. However, for the other minerals such as sillmanite, only the −395 μ particles would report to the overflow. Therefore the harder to melt and separate −600 +395 μ particles would report to the underflow as waste or to be used as a product where melting characteristics are not important.

The spiral, now separating feed that is mostly void of refractory heavy minerals in the coarser size range, can more efficiently separate the finer particles. Data from a glass sand plant in India showed that the feed contained 2–4 grains of refractory heavy minerals per kilogram of feed to the plant. The majority of these grains were in the coarser size fractions that would result in stone defects in the glass. At a plant feed rate of 40 t/h, this would result in a minimum potential for 80,000 glass defects per hour.

The process has shown to be very effective with little or no sillmanite of the size that could cause glass defects being in the final product after three years of operation. A photograph of the glass sand plant in Figure 3 shows the LC-3700 spirals and the Floatex® density separator.

FIGURE 3 St. Gobain glass sand plant

REFERENCES

Delgado, L.F. 1994. Spiral Applications in Mexico–Industrial Sand and Coal. SME Annual Meeting, Albuquerque, New Mexico, February 14–17.

Sadowski, Jim. 2000. Physical Separation Techniques for the Preparation of Glass Sand. *60th Conference of Glass Problems*. pp. 123–145.

Outokumpu Physical Separation. 2001. *Internal Document, Phase 1 Report*. November 2.

Vargas, Hugo. 2002. Email Data Transmittal from Del Istmo to Steve Hearn of Outokumpu Physical Separation. October 23.

Index

A
Ash content 84–85

C
Carpco spiral concentrators 182
Centrifugal jigs
  in fine coal cleaning 134
  at Granny Smith Gold Mine (Australia) 155–164
CI. *See* Cleanability index
Cleanability index, 86–90
Coal
  centrifugal jigs in fine coal cleaning 134
  dense medium cyclones in fine coal cleaning 55–58
  enhanced gravity separators in fine coal cleaning 133–137
  fine coal separation innovations 125–140
  flowing film in fine coal cleaning 134
  and HydroFloat separator 175–177
  optimum cutpoints for heavy media circuits 81–91
  role of magnetite in fine coal cleaning 58–70
  spiral concentrator circuits in fine coal cleaning 127–130
  teeter-bed separators in fine coal cleaning 130–133, 134–135
  water-only cyclone/spiral circuits in cleaning fine coal 93–106
CrossFlow teeter-bed separators 115–124
Curragh Mine (Australia) 56–58

D
Dense medium cyclones
  in fine coal cleaning 55–58
  modeling generalized partition curves for characterization of performance 71–78
  and role of magnetite in fine coal cleaning 58–70
Density-based separation processes
  plant control alternatives 9–12
  and process engineering principles 1–15
Dynamic hindered-settling model 39–53

E
Enhanced gravity separators 133–137
Ep. *See* Process efficiency (Ep)
Eriez 115

F
Feldspar and HydroFloat separator 174–175
Floatex density separators 182
Flowing film 134
Fluidized beds
  hindered settling of suspensions in 19–38
  inclined plates to increase segregation rates 31–35
  inversion in 22–23
  measurement of size and density distribution of particle feed 26–29
Froth flotation. *See* HydroFloat separator

G
Glass sand production with spiral concentrators 179–187
Gold 155–164
Granny Smith Gold Mine (Australia) 155–164
Gravity concentration. *See* Dense medium cyclones, Fluidized beds, Heavy media separation, Hindered settling, Teeter-bed separators
Greenside Colliery (South Africa) 56

H
Heavy media separation 143–144
  commercial applications, 148–149
  equipment, 147–148
  optimum cutpoints 81–91
  technology, 144–145, 149–153
  testing, 145–147
Hindered settling. *See also* Teeter-bed separators
  in fluidized beds 19–38
  modeling column separations 39–53
HMS. *See* Heavy media separation
HydroFloat separator 165–178

I
Incremental quality concept 2–4
  in analysis of two-stage spiral configurations 14
  in evaluation of plant control alternatives 9–12
  and heavy media cutpoints 82–84
Independence Coal Co. (West Virginia) case study of single-assembly two-stage spiral configuration 107–114

K
Kelsey centrifugal jig 155–164

L

Linear circuit analysis 4–7
   and two-stage spiral configurations 12–14

M

Magnetite
   bimodal particle size distribution 62–69
   particle size distribution 58–59
   rheology of dense medium 59–62
Materiales del Istmo (Mexico) 182
Meren (Jordan) 182–183
Modeling
   generalized partition curves for characterization of dense medium cyclone performance 71–78
   hindered settling column separations 39–53
   slip velocity 23–25
Multotec Process Equipment single-assembly two-stage spiral configuration 107–114

P

Partition curves 71–78
Phosphate and HydroFloat separator 169–173
Potash and HydroFloat separator 173–174
Process efficiency (Ep) 85–90

R

Reflux Classifiers 31–35

S

Silice del Istmo (Mexico) 182
Single-stage spiral configurations 128–129

Slip velocity
   measurement of 29–31
   modeling 23–25
Specific gravity and optimum cutpoints in coal cleaning 84–90
Spiral concentrators. *See also* Single-stage spiral configurations, Two-stage spiral configurations, Water-only cyclone/spiral circuits
   in glass sand production 179–187

T

Teeter-bed separators 115–119, 123–124. *See also* HydroFloat separator
   compared with spirals 119–120
   in fine coal cleaning 130–133, 134–135
   testing and full-scale installation 120–122
Two-stage circuits (water-only cyclone/spiral circuits in cleaning fine coal) 93–106
Two-stage spiral configurations 127–128
   and linear circuit analysis 12–14
   single-assembly two-stage spiral compared with double-stage spiral circuit 107–114

W

Water-only cyclone/spiral circuits
   with middlings recycle, 97–100
   operating characteristics in cleaning fine coal 93–106
   with no recycle, 101–102